Proceedings of the Fourth Workshop on

Science with the New Generation of High Energy Gamma-Ray Experiments

The Variable Gamma-Ray Sources: Their Identifications and Counterparts

Proceedings of the Fourth Workshop on

Science with the New Generation of High Energy Gamma-Ray Experiments

The Variable Gamma-Ray Sources: Their Identifications and Counterparts

Isola d'Elba, Italy 20 – 22 June 2006

editors

Marco Maria Massai
Università di Pisa & INFN Pisa, Italy

Nicola Omodei
INFN Pisa, Italy

Gloria Spandre
INFN Pisa, Italy

World Scientific

NEW JERSEY • LONDON • SINGAPORE • BEIJING • SHANGHAI • HONG KONG • TAIPEI • CHENNAI

Published by

World Scientific Publishing Co. Pte. Ltd.
5 Toh Tuck Link, Singapore 596224
USA office: 27 Warren Street, Suite 401-402, Hackensack, NJ 07601
UK office: 57 Shelton Street, Covent Garden, London WC2H 9HE

British Library Cataloguing-in-Publication Data
A catalogue record for this book is available from the British Library.

SCIENCE WITH THE NEW GENERATION OF HIGH ENERGY GAMMA-RAY EXPERIMENTS
The Variable Gamma-Ray Sources: Their Identifications and Counterparts
Proceedings of the Fourth Workshop

ISBN-13 978-981-270-964-6
ISBN-10 981-270-964-9

Printed in Singapore by World Scientific Printers (S) Pte Ltd

Fourth Workshop on:

Science with the New Generation of High Energy Gamma-ray Experiments
The variable gamma-ray sources: their identifications & counterparts

June 20 - 22, 200 Portoferraio, Elba Island (Italy)

Introduction

This is the fourth Workshop of a series on High Energy Gamma-ray Experiments, following the Conferences held in Perugia 2003, Bari 2004, and Cividale del Friuli 2005. The Workshop is organized by the Physics Department of Pisa University and by INFN which is actively participating in the new generation of high energy gamma-ray astrophysics experiments with GLAST, AGILE, MAGIC and ARGO and in the development of new technologies and preparation of new scientists.

The driver theme of this International Workshop is the study of the transient nature of the high-energy universe.

The high-energy gamma-ray sky presents a variety of sources that are deeply connected with the highest-energy particle acceleration mechanisms, in sites where intense magnetic fields, strong gravitational regimes and explosions of massive objects may take place. Some of these sources present a peculiar variability on different time scales depending on their origin. Some sources provide quasi-periodic signals, like Pulsars or Soft Gamma-Repeaters, some other are destructive as Gamma-Ray Bursts, while others are flaring as the Sun or Active Galactic Nuclei.

The Workshop want to address the topic of variable gamma-ray sources from the perspective of their identification and counterparts at different wavelengths and give an overview of the theory related to the most qualified emission processes that take place in these sources and of the nature of their variability.

To observe gamma-rays from distant objects in space, detectors need to be placed out the Earth's atmosphere using balloons or spacecrafts so the history of gamma-ray astronomy is closely tied to that of the space program in late 1960's. It was the Explorer XI satellite that, in 1961, first picked up fewer than 100 cosmic gamma-ray photons. Since that time many other missions have been realized to learn as much as possible of the gamma-ray sky. In 1972 the SAS-2 satellite discovered a diffuse gamma-ray background and the COS-B satellite (1975 - 1982) produced the first detailed map of the sky at gamma-ray wavelength, but it was the heaviest astrophysical payload ever flown before, the Compton Gamma-Ray Observatory (CGRO) that in April 1991, with its four major experiments aboard, covered an unprecedented six decades of the electromagnetic spectrum, from 30 keV to 30 GeV, and greatly improved the spatial and temporal resolution of gamma-ray observations. On the wake of the extraordinary discoveries of CGRO gamma-ray astronomy has attracted more and more interest and new sophisticated gamma-ray detectors on spacecraft have been designed and built. At the same time large and sensitive ground-based telescopes, the only one that can the TeV region of the gamma energy spectrum, have been put in operation.

Gloria Spandre
Nicola Omodei

CONTENTS

I Space-based Telescope

INTEGRAL: 4 YEARS IN ORBIT

PIETRO UBERTINI
IASF-INAF, Via Fosso del Cavaliere100, 00133 Roma- ITALY

PATRIZIA CARAVEO
IASF-INAF, Via Bassini, 15- 20133 Milano- ITALY

After 4 years of operation, ESA's INTEGRAL mission has produced a remarkable portfolio of results, ranging from the inventory of the high energy sources, to the discovery of dozens of variable sources to the mapping of the Al and annihilation line in the Galaxy. A brief review of the most dramatic achievements of the mission is presented.

1. Introduction

The ESA observatory INTEGRAL (International Gamma-Ray Astrophysics Laboratory) [1] is dedicated to the fine spectroscopy (2.5 keV FWHM @ 1 MeV) and fine imaging (angular resolution: 12 arcmin FWHM) of celestial gamma-ray sources in the energy range 15 keV to 10 MeV. While the imaging is carried out by the imager IBIS [2], the fine spectroscopy is performed by the spectrometer SPI [3]. In addition, source monitoring in the X-ray (3-35 keV) and optical (V-band, 550 nm) energy ranges in provided by the JEM X and OMC instruments [4,5].

The spectrometer, imager and X-ray monitor are based on the use of coded aperture mask. Such a technique is the key for providing images at energy above tens of KeV, where photons focussing become impossible using standard grazing technique. Moreover, coded mask eases background subtraction because, for any particular source direction, the detector pixels are split into two intermingled subsets, those capable of viewing the source and those for which the flux is blocked by opaque mask elements. Thus, the subset which is not directly illuminated by the source provides contemporaneous background measurement for the illuminated one, under identical conditions.

The technique, discussed in detail by [6] is extremely effective in controlling the systematic errors and has achieved remarkable results in crowded galactic regions, such as the very center of our Galaxy.

INTEGRAL was launched from Baikonur on Oct.17th, 2002. It was inserted into an highly eccentric orbit (characterized by a 9,000 km perigee and by a 154,000 km apogee) providing long periods of uninterrupted observation with nearly constant background and away from trapped radiation (electron and proton radiation belts). Owing to background radiation effect in the high-energy detectors, scientific observations are carried out while the satellite is above a nominal altitude of 60,000 km (approaching radiation belts) and above 40,000 km (leaving radiation belts).
This means that about 90% of the time can be used for scientific observations. The data are received in realtime by the INTEGRAL Science Data Center (ISDC) [7] which prepares the final data products to be distributed to the observes and later archived for public use.

2. The source catalogues (and consequences).

Using 40 Ms exposure time, i.e. all the IBIS data available up the and of May 2006, Bird et al [8] have compiled the 3rd INTEGRAL source catalogue which encompasses 421 sources, detected above 4.5 σ in the energy range 18-100 keV.

Since INTEGRAL is frequently observing in the Galactic Plane, the sky coverage is far for uniform (see fig.1 of [8]). Thus, it does not come as a surprise that the majority of the IBIS sources is located at low galactic latitudes. Irrespective of source location, the identification process is based on a multiwavelength approach, taking advantage of Radio, IR and X-ray archival data. Ad hoc optical and IR observing campaigns are also actively pursued [9].

Considering only the firm identifications, 171 sources (i.e. 41%) have been associated with galactic accreting systems, 122 (i.e. 29%) with extragalactic objects, 15 with different classes of celestial emitters, while 113 (i.e. 26%) are still awaiting identification.

The galactic identifications can be divided into 21 CV systems (9 of which are new detections with emission extending up to 100 keV), 65 high-mass X-ray binaries (HMXB) and 78 low-mass X-ray binaries (LMXB). We note that while INTEGRAL continues to detect LMXB, the rate of discovery is much lower than for the high mass systems. In particular, the HMXB sample encompasses also 19 new INTEGRAL sources which have been identified with Be binary systems on the basis of their spectral characteristic and/or transient behavior. Indeed, Bird et al [8] point out the emergence of a new class of supergiant fast X-ray transient SFXT). The efforts to identify new variable sources discovered by INTEGRAL (known as IGR Jxxxxx.yyyy) resulted in 8 firm identifications and 4 probable ones.

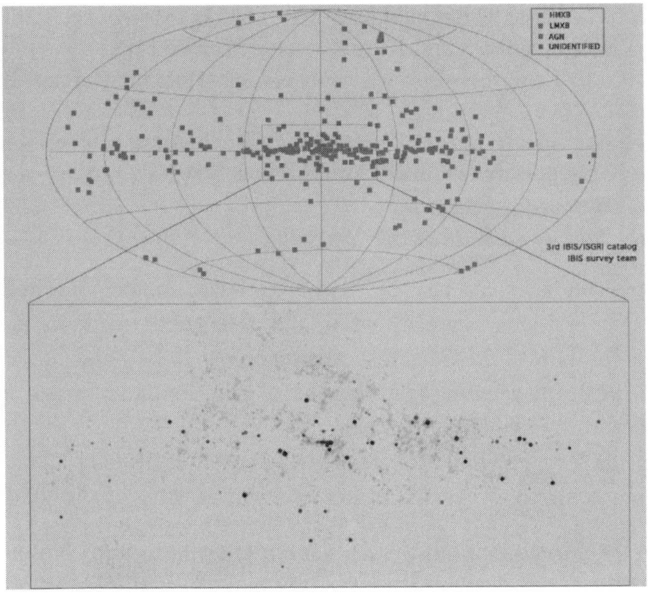

Figure 1: The upper image shows the distribution on the sky of four of the main soft gamma-ray source populations observed in the third INTEGRAL/IBIS survey catalogue. This newly-released catalog contains 421 sources. Of the known systems, the low-mass X-ray binaries (LMXB) are old systems mainly populating the galactic bulge, the high-mass X-ray binaries (HMXB) are younger systems seen along the galactic plane, and the active galactic nuclei (AGN) are extragalactic sources seen over the whole sky. Around one in four of the sources seen by INTEGRAL are unidentified, and their distribution is also shown.
The lower image shows a false colour image of the central region of our galaxy (+40°<l<-40°, -18°<b<+18°). (Credit: IBIS Survey Team)

2.1. *The sky above 100 keV*

Selecting energies above 100 keV, Bazzano et al [10] have provided the first catalogue of very hard IBIS sources.

While 49 sources are detected with a significance above 4σ in the 100-150 keV energy range, only 14 of them are seen above 150 keV.
Accreting galactic systems dominate the high energy catalogue with 26 LMXBs, 6 HMXBs , 2 anomalous X-ray pulsars and 1 unidentified source, thought to be a Black Hole candidate in a LMXB. Moreover 1 Soft Gamma Repeater, 3 isolated pulsars and 10 AGNs have been detected.
Above 150 keV, on the contrary, only Black Hole candidates have been detected.
Considering the source spectral shape, the two hardest sources are the two anomalous X-ray pulsars, while the softest one is ScoX-1.

2.2. *On the lack of galactic diffuse emission*

Source catalogues can also be used to estimate source contribution to the total galactic emission. Indeed, while the interstellar medium is probably responsible for most of the Galactic emission below 10 keV and above 300 keV, below 300 keV, the relative contributions of compact sources and interstellar emission processes are highly uncertain.

Using the first INTEGRAL catalogue, Lebrun et al. [11] found that the point source contribution accounts for as much as 90% of the total emission from the Galaxy, leaving no more than 10-15% to diffuse processes. It is important to note that, while the data do not rule out the presence of diffuse processes, the total source contribution could easily account for the totality of the galactic emission in soft gamma-rays.

A similar result is reported by Bazzano et al. [10] using the sources detected above 100 keV. Thus, point sources dominate the galactic emission in the energy domain covered by the IBIS instrument.

3. Using the Earth to measure diffuse background

Assessing the extragalactic diffuse background is a daunting task for every astronomical mission. To allow for a precise measurement of the diffuse emission in the energy range 5-100 keV, INTEGRAL has been operated in an unconventional way, using the Earth as a shield to eclipse part of the field of view. Combining the decrease of cosmic gamma ray flux when the Earth blocked the Observatory field of view together with the accurate calculation of the contribution due to the emission of the earth atmosphere has yielded data of exceptional accuracy on both the intensity and the spectral shape of the X and Gamma background [12] probably due to distant and weak objects that cannot be detected individually.

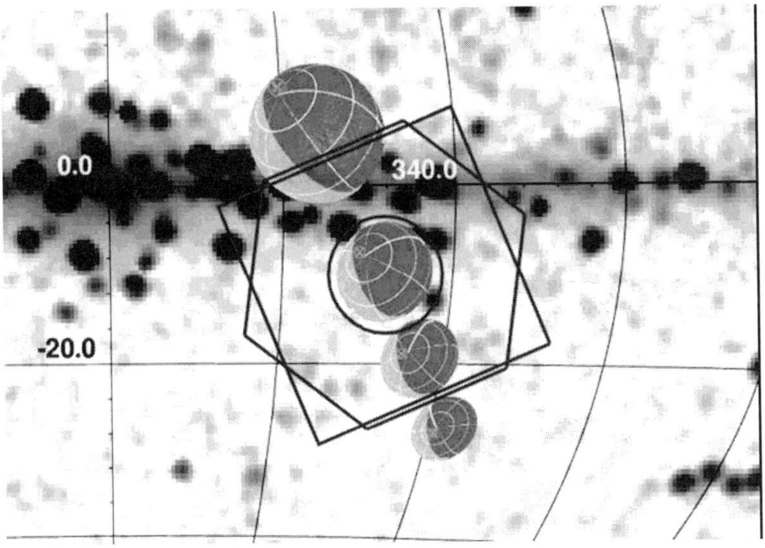

Figure 2: Schematic view of the INTEGRAL Earth observing mode. The zero sensitivity Field of Views (FoVs) of JEM-X (circle), IBIS (square) and SPI (hexagon) are shown superposed on to the RXTE 3-20 keV slew map.
The linear sizes of the fully coded FoVs correspond to half of the area shown.
In this map, beside a large number of compact sources, an extended X-ray emission associated with the Galactic Ridge is also visible. In Earth observing mode the pointing direction of the telescopes remains the same, while the Earth crosses the instruments FoVs. Earth day-side is shown by lighter shade of grey. During the observation the Earth moves from positive to negative latitudes. Since the distance from the Earth increases rapidly during this part of the 3-day INTEGRAL orbit, the angular size of the Earth disk decreases [12].

4. Gamma Ray Bursts

Since forty years astronomical satellites are recording once a day intense flashes of gamma radiation (Gamma ray bursts, GRB), apparently coming from random sky directions. GRB can last few thousandths of a seconds till over an hour. They are, for a short time, the brightest objects in the sky and, apart from a well defined subclass, they never repeat.
Although GRBs are not amongst the primary scientific objectives of the INTEGRAL mission, owing to the relatively big field of view of the IBIS and SPI instruments, the probability to detect GRBs is not negligible. Indeed, INTEGRAL does observe about one Gamma burst each month and, thanks to

the real time connection to the ISDC, is possible to compute in real time accurate positions and circulate the information within few seconds to the international scientific community[13].

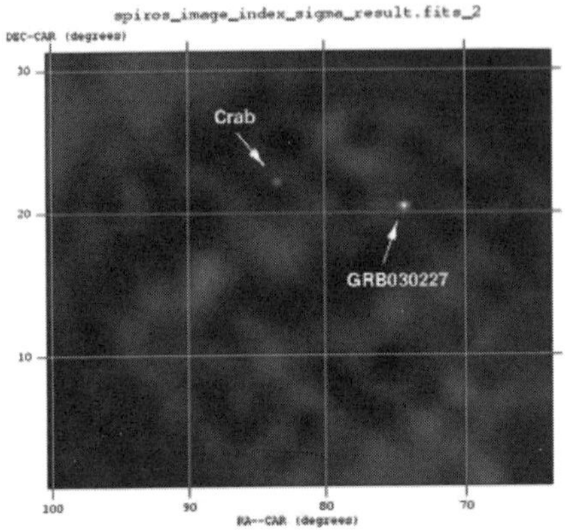

Figure 3: GRB030227: a strong GRB seen by SPI in the vicinity of the Crab.

5. The radioactive galaxy

The INTEGRAL spectrometer SPI focussed on the observation of the galactic plane to have an accurate measurement of the nuclear emission caused by the decay of Al 26 (1809 keV) produced by supernovae exploded in the last million years. The extremely good spectral resolution of SPI allowed for the measurement of the "Doppler" shift of the Al 26 emission line, due to the Milky Way rotation. While the radiation, coming from the Milky Way central area, does not show any noticeable shift, this is not true for the radiation coming from the disk, which rotate with the whole Galaxy [14].
Such measurement allowed also to estimate that the rate of the core-collapse supernovae is between 1 and 3 per century.

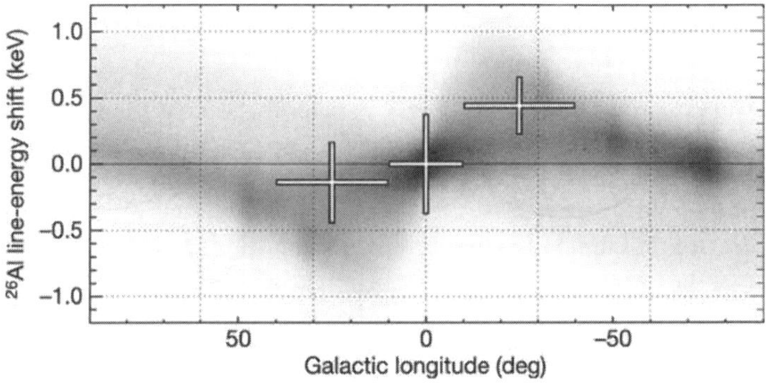

Figure 4: A detailed modeling of the Doppler shift of the line energy (1809 keV) measured by SPI is consistent with the expected one from the galactic rotation. The map shows this expectation based on modeling the Galactic rotation curve and a 3-dimensional distribution of ^{26}Al sources. The observed line shift is less than 0.5 keV, demonstrating the superb quality of SPI's spectral resolution capabilities. The dominant sources of ^{26}Al emission are massive stars. This observation shows that ^{26}Al emission is truly global, as we see the entire amount of it throughout our Galaxy and co-rotate with the Galaxy [14].

The situation is completely different when imaging the 511 keV line [15]. In spite of a deep coverage of the galactic plane, the 511 keV emission is concentrated in the galactic center where is detected at the 50σ level, while a much weaker emission, at 4σ level, is detected from the galactic plane . The bulge emission is highly symmetric and is centered on the galactic center with an extension of 8° (*FWHM*). No evidence for a point-like source in addition to the diffuse emission was found, down to a typical flux limit of $\sim 10^{-4}$ ph cm^{-2} s^{-1}.

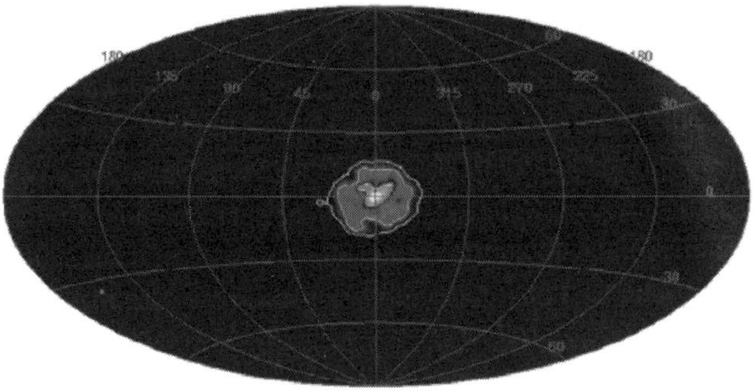

Figure 5: Richardson-Lucy image of 511 keV gamma-ray line emission from [15]. Contour levels indicate intensity levels of 10^{-2}, 10^{-3}, and 10^{-4} ph cm^{-2} s^{-1} sr^{-1} (from the centre outwards).

6. Conclusions

After 4 years in orbit with all the instruments performing well, the INTEGRAL mission has been extended up to 2010. Deeper and more even exposure of the whole sky will be achieved allowing to reach a limiting flux of 1 mCrab, thus increasing dramatically the number of sources detectable

Acknowledgments

This research has been supported by ASI under contract I/R/046/04. The authors thank the IBIS Survey Team and all their collaborators for continuum effort and enthusiasm.
The data are based on observations performed by INTEGRAL, an ESA Project with instruments and science data centre funded by ESA member states (especially the PI countries: Denmark, France, Germany, Italy, Switzerland, Spain), Czech Republic and Poland, and with the participation of Russia and the USA.

References

1) C. Winkler et al, A&A, 411, L1 (2003)
2) P. Ubertini et al. A&A, 411, L131 (2003)
3) G. Vedrenne et al. A&A, 411, L63 (2003)
4) N. Lund et al. A&A, 411, L231 (2003)
5) M. Mass-Hesse et al. A&A , 411 L261 (2003)
6) G. Skinner & P.Connell, A&A, 411, L123 (2003)
7) T. Courvoisier et al. A&A, 411 L49 (2003)
8) A.J. Bird et al. Ap.J. Lett in press (2007)
9) N. Masetti, et al., A&A, 459, 21 (2006)
10) A. Bazzano et al. Ap.J. 649, L9 (2006)
11) F. Lebrun et al. Nature, 428, 293 (2004).
12) E. Churazov, et al., A&A, (2007, in press)
13) S. Mereghetti et al, A&A, 411, L291 (2003)
14) R. Diehl et al. Nature, 439, 45 (2006)
15) J. Knodlseder et al. A&A 441, 513 (2005)

The Suzaku Mission

Kazutaka Yamaoka on behalf of the Suzaku team

Department of Physics and Mathematics, Aoyama Gakuin University
Fuchinobe 5-10-1, Samamihara, Kanagawa 229-8558, Japan
**E-mail: yamaoka@phys.aoyama.ac.jp*

Suzaku satellite, the Japan-US collaborative mission, was successfully launched on July 10, 2006. It is equipped with 5 soft X-ray telescopes (XRT), one micro-calorimeter (XRS), 4 CCD cameras (XIS, and one hard X-ray detector (HXD). Though XRS is not operational, XIS and HXD provide us with new views of thermal and non-thermal phenomena. After the performance verification phase with 70 targets, the first term of general observer program (AO-1) has been started from April 1.

1. Suzaku Satellite

1.1. *mission overview*

The fifth Japanese-US X-ray satellite Astro-E2 was successfully launched by M-V rocket from Uchinoura Space Center on July 10, 2005. It was renamed as "Suzaku" satellite after red Chinese phoenix [1], then thrown into a near circular orbit with 550 km altitude with an inclination angle of 31 degree as schedule. Suzaku carries five soft X-ray instruments on each focal plane of five X-ray Telescopes (XRT [2]) and one hard X-ray detector (HXD [3] [4]). There are two types of soft X-ray detectors covering 0.2–12 keV. One is micro-calorimeter array (XRS [5]) with superior energy resolution of 6 eV@6 keV, developed by NASA/GSFC and ISAS/JAXA. The XRS achieved the best energy resolution of 7 eV@ 6 keV in space for the first time, but it had an accident that all the liquid helium to cool the detector was evaporated on August 8, hence, the XRS is not operational at present. The other soft X-ray detector is four X-ray CCD cameras (XIS [6]) which are similar types of ASCA SIS. One of the XISs, XIS1, uses a back-side illuminated CCD (BI), while the other three use front-side illuminated CCDs (FI). The HXD is a combination of GSO/BGO well-type phoswich scintillator and Si PIN diodes. It has been designed to have very low backgrounds relative to small detection area. In addition to these two narrow-field instruments, the lateral anti-coincidence shields of the HXD have been designed to work as a wide-field instrument covering about half the sky with an energy range of 50–5000 keV. This is called wide-band all-sky monitor (WAM [7]) to monitor transient phenomena such as gamma-ray bursts, hard X-ray transients, and solar flares. The characteristics of all the Suzaku instruments are summarized in Table 1.

Table 1. Characteristics of the Suzaku instruments [1].

	XRT + XIS FI/BI	HXD PIN/GSO	HXD-WAM
number of detector	3/1	1/1 (16 units)	4 (20 units)
point spread func.	2' (HPD)	–	–
energy range (keV)	0.2–12	10–70/40–600	50–5000
energy resolution	~130eV@6keV	~3keV/7.6$\sqrt{E_{\rm MeV}}$%	30%@662keV
effective area (cm^2)	330/370@1.5keV	160@20keV/260@100keV	400@1 MeV
field of view	17.8'×17.8'	34'×34'(<100keV) 4.5°×4.5°(>100keV)	~2π
time resolution	8 s, 7.8 ms (P-Sum)	61 μs	1 s/15.6 ms (GRB data)

1.2. *Advantage of Suzaku*

The XIS/BI has an excellent line-response below 1 keV with little tail component. For monochromatic oxygen K_α lines with 0.65 keV, the FWHM energy resolution is ~65 eV. The Suzaku XRT+XIS also has a large collecting area of ~ 1000 cm^2 in total, although the spatial resolution (Half Power Diameter 2') is poorer than Chandra and XMM-Newton. At iron-K band around 6–7 keV, the Suzaku effective area is comparable to that of XMM-Newton PN. All four XISs have low backgrounds, due to a combination of the Suzaku orbit and the instrumental design. The large effective area at Fe K combined with this low background make Suzaku a powerful tool for investigating hot and/or high energy sources as well.

The HXD has a good sensitivity in 10–600 keV due to its low instrumental background by its tight active shielding and narrow field of view. Figure 1 shows the HXD non-Xray background in comparison with Beppo-SAX/PDS and RXTE/HEXTE. The HXD achieved the lowest non-Xray background in the energy range of 12–70 keV and 150–500 keV. The HXD does not take the background by changing the field of view, hence the sensitivity is determined by not only statistical fluctuation of the background but also reproducibility of background modeling. The current background model gives an accuracy of reproducibility of 5% and 3% for PIN and GSO respectively, corresponding to the detection limit of ~0.5 mCrab in the

PIN energy range and ~10 mCrab in the GSO energy range for 100 ksec exposure. Our goal is aimed at ~ 1% level.

In summary, Suzaku can provide excellent capability in the following observations.

- Plasma diagnostics of diffuse sources.
- Search for X-ray counterpart of diffuse GeV/TeV emissions.
- Detailed study of iron-K band of AGNs and X-ray binaries.
- Wide-band spectroscopy in 0.2–600 keV to investigate connection between thermal and non-thermal component, i.e. to search for hard tail and high energy cutoff.
- Variability study of AGNs and X-ray binaries in a short time-scale of a few hours.

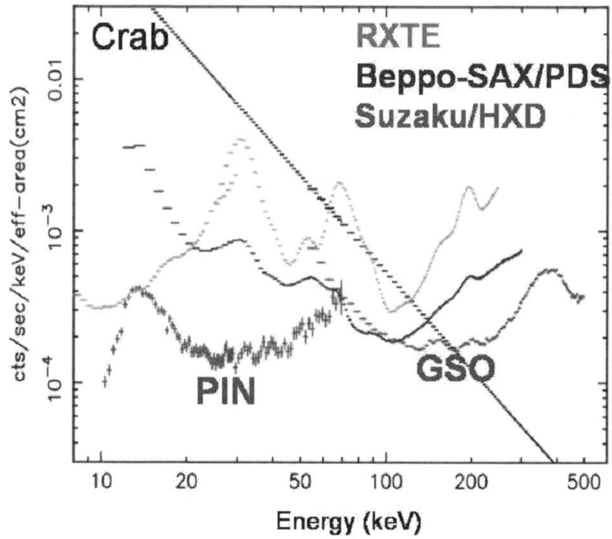

Fig. 1. Normalized HXD background in comparison with those of other hard X-ray instruments: Beppo-SAX/PDS and RXTE/HEXTE [4].

2. Initial results from Narrow Field Instruments

We have observed 70 sources during the first half-year operations with Suzaku. These targets were selected by the Suzaku Science Working Group (SWG) from four categories (Galactic/extra-galactic compact and diffuse). In order to show Suzaku excellent performance, we will present some results in this section.

The XIS took the first light image and spectra of supernova remnant (SNR) E0102.2-729 on August 14, 2006. Figure 2 shows the energy spectrum taken by XIS/FI and BI chips [6]. Owing to the excellent line response and high quantum efficiency in lower energy than 1 keV of the XIS/BI chip, Kα lines from highly ionized O at 0.57 keV and 0.65 keV are clearly resolved. Other samples are clear detections of K_α lines due to highly ionized irons as well as iron-K_β and nickel K_α lines in the Galactic center [8], detection of emission-lines from light element C, N and O in the planetary nebula BD +30 3639 [9] and the supernova remnant Cygnus loop [10], and so on. Thus, better efficiency and energy resolution of CCD is suitable to investigate emission lines from light element C and O, as well as Fe-K lines. Objectives are planetary nebula, supernova remnants, Galactic center and cluster of galaxies. The status and origin of the plasmas are well examined from the line energies, line ratios, and line broadening.

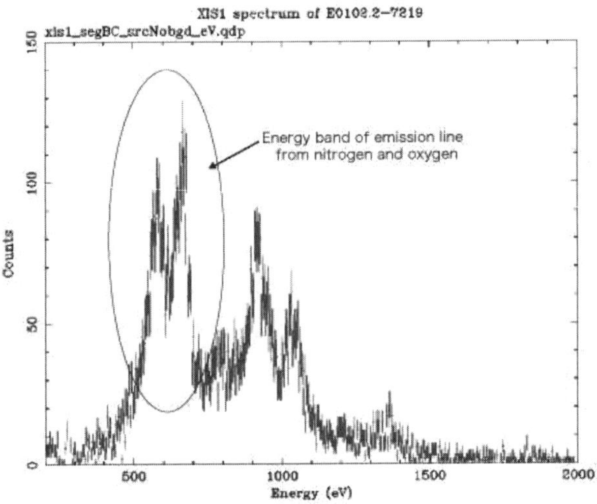

Fig. 2. XIS first-light spectrum of the supernova remnant E0102.2-729 [6].

The HXD made its first observation of the radio galaxy Centaurus A (NGC 5128) on August 18, 2006. The spectrum was detected up to 200 keV, covering the energy range of 2–200 keV. Thus, another area, Suzaku satellite has advantages, is broadband spectroscopy with better sensitivity. Targets are X-ray binaries, active galactic nuclei, and cluster of galaxies. Figure 3 shows the wide-band spectrum of the blackhole candidates Cygnus X–1 [11] and GRO J1655–40 [12] in the low/hard state. For such bright sources with an intensity level of 100 mCrab, the HXD can detect it up to 300 keV or more. All the spectral components (power law-like intrinsic emissions, soft excess, reflection component, and iron-K lines) appear in the spectrum, which enable us to study accretion geometry around the black hole. Hard tail in the supernova remnant and cluster of galaxies is very important in clarifying the relation between non-thermal and thermal component. In addition to confirmation of a presence of hard tail from Casiopea A, hard X-rays from the TeV SNR RXJ1713.7-3946 are clearly detected up to 40 keV for the first time [13]. Search for the hard tail from other SNRs and galaxy clusters is still in progress.

We have observed the Seyfert-1 galaxy MCG-6-30-15 which is famous to show broad iron-line emissions for a long exposure of 300 ksec [14]. They could be considered as direct evidence for emissions from spinning black hole. Suzaku confirmed broad iron lines with accurate determination of underlying continuum components. Figure 4 shows broadband spectrum in the high and low-flux phase of the light curve. Iron-K emissions and the reflection component around 20 keV shows almost constant flux level, while the soft component is variable. It can be interpreted as a light bending effect around the black hole. Thus, Suzaku is powerful in searching for time-variability of hard X-rays in a short time scale of ~hours. It was revealed that the Seyfert 1.9 galaxy MCG-5-23-16 showed also broad-line features and less variability in the hard X-ray band [15].

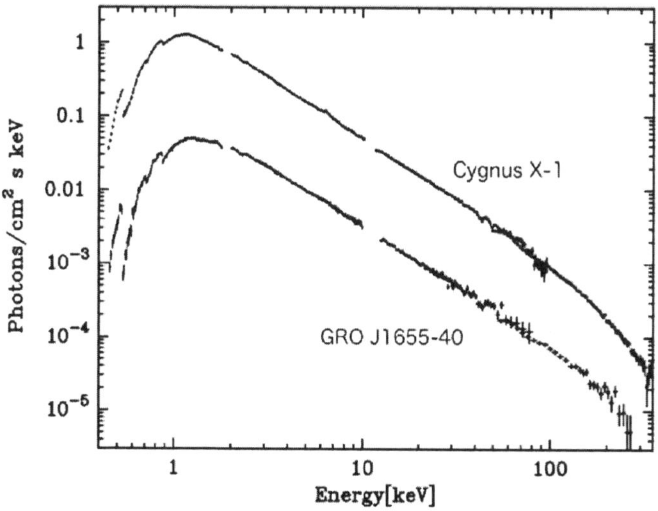

Fig. 3. Suzaku broadband spectra of the blackhole candidates Cygnus X-1 and GRO J1655–40 [11] [12].

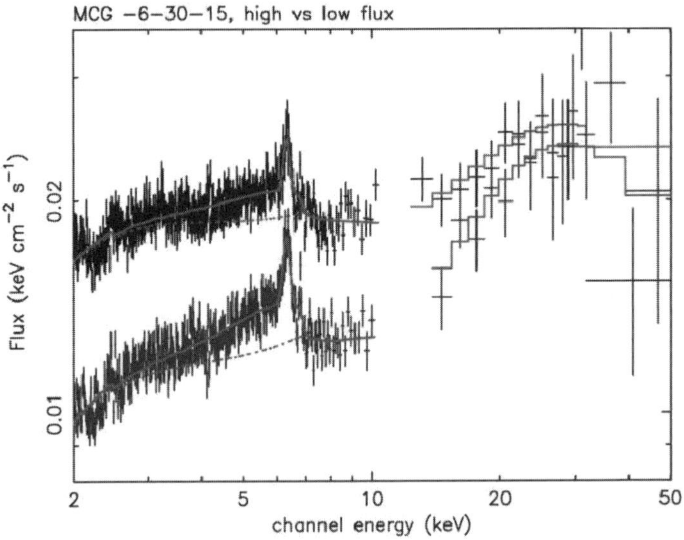

Fig. 4. Suzaku multi-band spectra of high- and low-flux phase of the Seyfert 1.5 Galaxy MCG-6-30-15 [14].

3. Suzaku Wide-band All-sky Monitor(WAM)

The Suzaku WAM is a subsystem of the HXD, utilizing 20 BGO anti-coincidence detector units. The WAM consists of four identical walls with each geometrical area of 800 cm^2. Figure 5 shows the on-axis effective area of one WAM detector in comparison with previous and future gamma-ray burst instruments. Thanks to its large geometrical area and high-Z material, it has a very large effective area, 400 cm^2, even at 1 MeV. Hence, the WAM could provide unique opportunities to detect high energy emissions from GRBs in the MeV range.

Since the WAM GRB function was activated on August 22 2005, we have detected more than 80 GRB candidates in first 8-months operations. This corresponds to a GRB detection rate of more than 120 per year. The GRB light curves show clear bimodal distribution with short and long durations as seen in the previous BATSE results. All the light curves for GRBs confirmed by other satellites are available at the following web page: http://www.astro.isas.jaxa.jp/suzaku/HXD-WAM/WAM-GRB/grb/grb_table.html.

The WAM has powerful spectral capabilities for well-localized GRBs

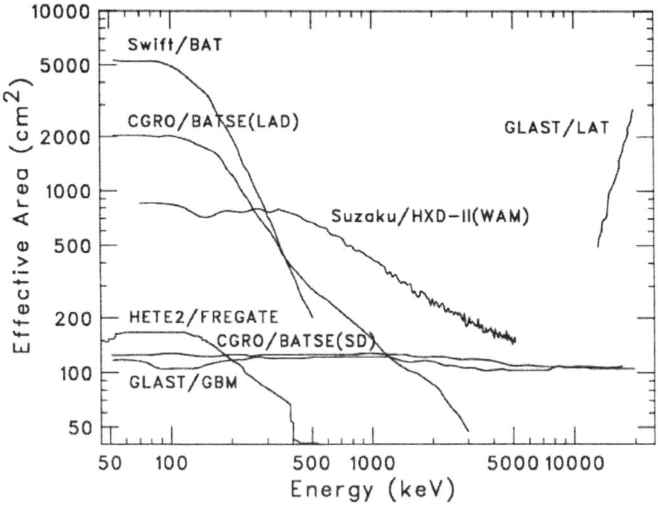

Fig. 5. HXD/WAM effective area in comparison with other GRB instruments [7].

by Swift/BAT, INTEGRAL/IBIS and inter-planetary network(IPN). Figure 6 shows the unfolded $\nu F\nu$ energy spectra of three Swift-GRBs: GRB 050915b, GRB 051111, and GRB 051008. It is impressive that the photons from the bright, hard GRB 051008 was detected up at least 1 MeV by the WAM. These GRBs were observed simultaneously with Swift/BAT, and we performed joint spectral analysis with Swift/BAT (15–150 keV) and WAM (100–2000 keV) over 15–2000 keV. The peaks of the spectra (Epeak) are clearly found in the 50 keV to 1 MeV range. The WAM in-flight calibration is still on-going, but such capability allows us to study some correlation in the prompt emissions. *e.x.* Amati-relation [16] and Ghirlanda-relation [17]. In fact, spectral parameters in GRB 051111 were verified to satisfy the Amati relation. Future AGILE and GLAST collaborations with GeV gamma-rays will be also very interesting.

4. Summary

The X-ray satellite Suzaku is now operating well with two instruments: XIS and HXD. The Suzaku power is 1)simultaneous wide-band coverage, 2)excellent soft X-ray response with better sensitivity, and 3)wide-band all-sky monitor(WAM) as a monitor of GRB prompt emissions. The exciting results from first-year operations will be presented in PASJ special issue and at the Suzaku conference held in Kyoto on December 2006. Suzaku Guest Observer Program (AO1) has started since April 1 2006. The AO2

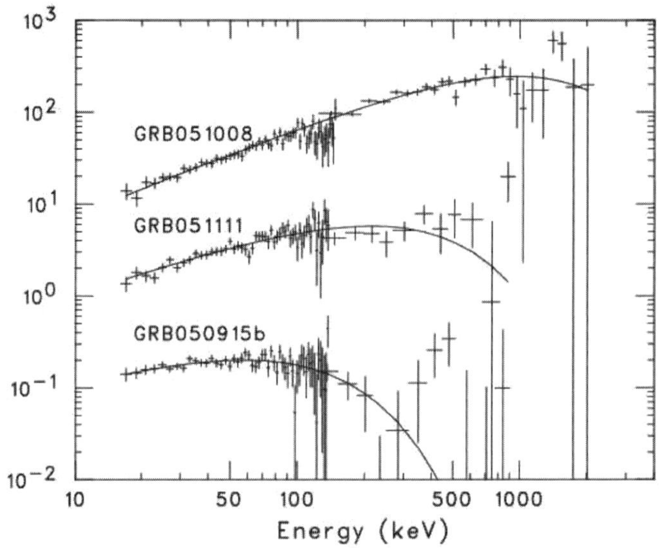

Fig. 6. Combined Swift/BAT and Suzaku/WAM spectra of the three GRBs (GRB 060915B, GRB 061008 and GRB 061111) [7].

will be started from April 1 2007.

We would like to thank for all the members of Suzaku mission. The Suzaku mission is supported by NASA/GSFC and JAXA/ISAS.

References

1. Mitsuda, K. et al., *PASJ* Suzaku Special Issue **59** in press (2006).
2. Serlemitsos, P. J. et al., *PASJ* Suzaku Special Issue **59** in press (2006).
3. Takahashi, T. et al., *PASJ* Suzaku Special Issue **59** in press (2006).
4. Kokubun, M. et al., *PASJ* Suzaku Special Issue **59** in press (2006).
5. Kelly, R. L. et al., *PASJ* Suzaku Special Issue **59** in press (2006).
6. Koyama, K. et al., *PASJ* Suzazku Special Issue **59** in press (2006).
7. Yamaoka, K. et al., *SPIE* **6266** 122 (2006).
8. Koyama, K. et al., *PASJ* Suzaku Special Issue **59** in press (2006).
9. Murashima, M. et al., *ApJL* **647** 131 (2006)
10. Tsunemi, H. et al., *PASJ* Suzaku Special Issue**59** in press (2006)
11. Kobota, A. et al., in preparation.
12. Takahashi, H. et al., *PASJ* submitted (2006).
13. Tanaka, T. et al., in preparation.
14. Miniutti, G. et al., *PASJ* Suzaku Special Issue **59** in press (2006)
15. Reeves, J. N. et al., *PASJ* Suzaku Special Issue **59** in press (2006).
16. Amati, L. et al., *MNRAS* **372**, 233-245 (2006).
17. Ghirlanda, G. et al., *ApJ* **616**, 331-338 (2004).

The Swift Mission
Two years of operation

A. Moretti*

INAF- Osservatorio Astronomico di Brera, via E.Bianchi, 46
Merate (LC), 23807, ITALY
**E-mail: alberto.moretti@brera.inaf.it*
www.brera.inaf.it

The *Swift* Gamma-ray Burst Explorer mission, launched on 2004 November 20, is a multiwavelength observatory for gamma-ray burst astronomy. Here we report on the status of the mission at 2 years from the launch. We use the exceptional dataset of GRB 060814 as an example to illustrate the kind of temporal and spectral analysis that *Swift* data allow.

1. The *Swift* Satellite

The *Swift* satellite [1] is a mission dedicated to the study of gamma-ray bursts (GRBs) and their afterglows. GRBs are detected and localized by the Burst Alert Telescope [2], and followed up at X-ray (0.2–10 keV) and optical/ultraviolet (1700–6000 Å) wavelengths by the X-Ray Telescope [3] and the Ultraviolet/Optical Telescope [4]. This innovative combination of gamma-ray instrument discovery, followed by an automated spacecraft response, allows new GRBs to be localized and studied at X-ray through optical wavelengths starting only seconds after the burst occurs. Since afterglows fade quickly, Swift's rapid ~ 1 minute response allows observations when the emissions are orders of magnitude brighter than the previous few-hour response capabilities (SAX, XMM, Chandra). *Swift* has so far (November 2006) observed some 200 GRB afterglows: this rich dataset has allowed to study in detail the X-ray light curves, both at early and late times, leading to the discovery of a complex behaviour. Coupled with optical data, either from UVOT or ground-based observatories, this has opened a new era in the afterglow modeling. The three instruments, are shown on the spacecraft in Figure 1.

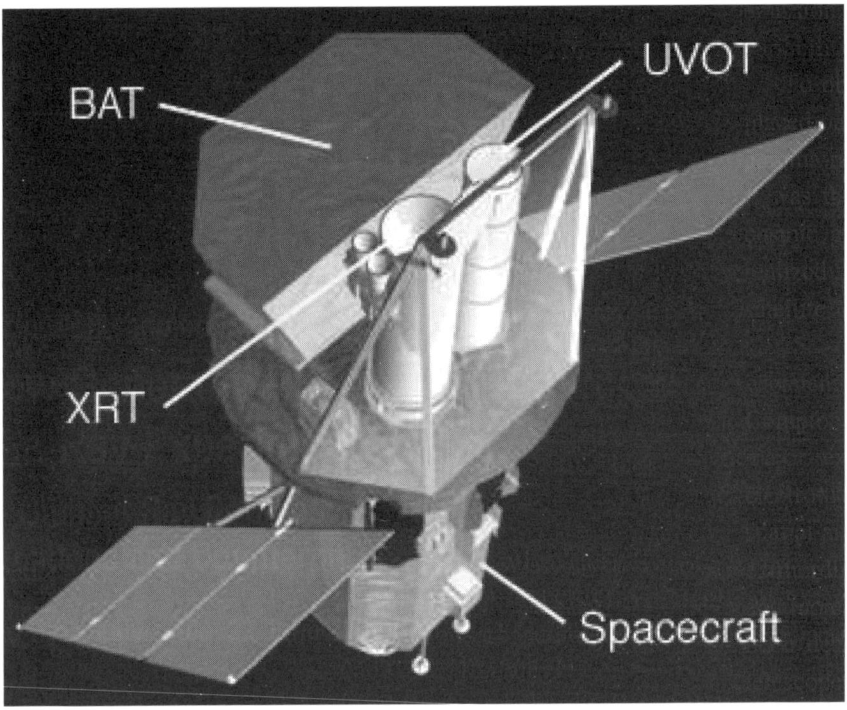

Fig. 1. The *Swift* satellite with its scientific payload.

The Burst Alert Telescope (BAT) is a highly sensitive, large FOV instrument designed to provide critical GRB triggers and 2-arcmin positions. It is a coded-mask instrument with a 1.4 steradian field-of-view (half coded). The energy range is 15-150 keV for imaging with a non-coded response up to 500 keV. Within the first $\sim$ 10 seconds of detecting a burst, the BAT calculates an initial position, decides whether the burst merits a spacecraft slew and, if worthy, sends the position to the spacecraft. Further information on the BATT is given by [2].

Swift's X-Ray Telescope (XRT) is a focusing X-ray telescope with a 110 cm^2 effective area (at 1.5 keV), 23 arcmin FOV, 18 arcsec resolution (half-power diameter at 1.5 krV), and 0.2-10 keV energy range. XRT is designed to measure the fluxes, spectra, and lightcurves of GRBs and afterglows over a wide dynamic range covering more than 7 orders of magnitude in flux. The XRT pinpoints GRBs to 5-arcsec accuracy within 10 seconds of target acquisition for a typical GRB and studies the X-ray counterparts of GRBs beginning 50-200 seconds from burst discovery and continuing for days to

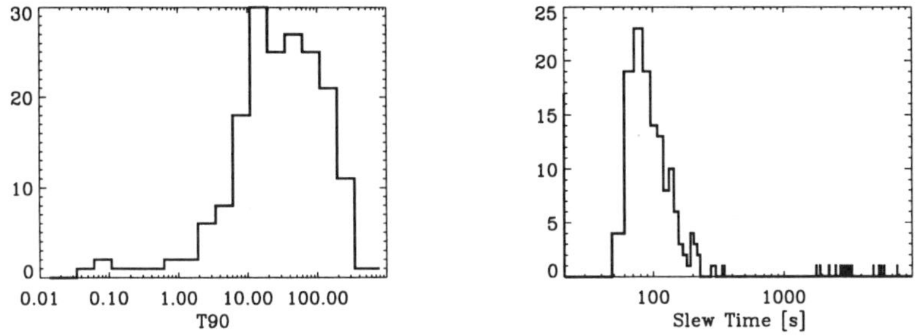

Fig. 2. **Left:** The T90 distribution of the BAT detected GRBs. **Right:** The slew time distribution

weeks. Further information on the XRT is given by [3].

The Ultra-Violet/Optical Telescope (UVOT) design is based on the Optical Monitor (OM) on-board ESA's XMM-Newton mission. UVOT is co-aligned with the XRT and carries an 11-position filter wheel, which allows low-resolution grism spectra of bright GRBs, and broadband UV/visible photometry. There is also a 4× field expander (magnifier) that delivers diffraction limited sampling of the central portion of the telescope FOV. Photons register on the microchannel plate intensified CCD (MIC). Further information on the UVOT is given by [4].

The launch was on a Delta 7320-10 on 2004 November 20 to a 22 degrees inclination, 600-km altitude orbit. Normal data are downlinked in several passes each day over the Italian Space Agency (ASI) ground station at Malindi, Kenya. Tracking and Data Relay Satellite System (TDRSS) are used to send burst alert messages to the ground. Similarly, information about bursts observed by other spacecraft are uplinked through TDRSS for evaluation by Swift's on-board figure-of-merit (FoM) software. There is no on-board propulsion system. Pointing is provided by momentum wheels with momentum unloaded by magnetic torquers. Pointing knowledge is through the gyroscopes, star trackers, sun sensors, and magnetometers. The observation strategy is for the XRT and UVOT to be almost constantly observing positions of bursts previously detected by the BAT. The most recently detected burst have priority (although this can be adjusted as science dictates). Typically 5 sources are observed each orbit. When a burst is detected, an automated series of XRT and UVOT observations are performed lasting 20,000 s in exposure. Following that time, additional observations are scheduled via ground planning. All three instruments are functioning

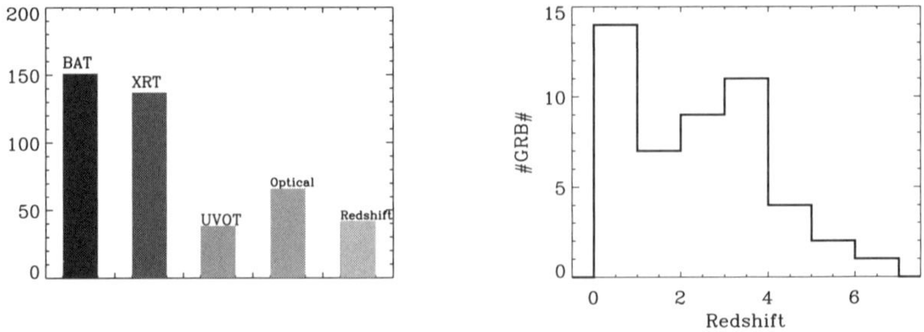

Fig. 3. **Left:** GRB and afterglow detection statistics. **Right:** The BAT detected GRB redshift distribution

nominally. The XRT has suffered a hit by a micro-meteoroid, but the effect of that damage has been mitigated by flight software changes. The XRT Thermo-Electric Cooler failed early in the mission, but operational changes have allowed XRT to operate on only passive thermal cooling. The UVOT has a filter wheel, which is a moving part, but there is a backup assembly in case of failure by the current wheel. For the details on the organization of the mission and on the observation status and efficiency see [5]

2. GRB sample statistics

The BAT first light was on the 3rd of December 2004, XRT first light was on the 11th of December 2004 and UVOT first light was on the 12th of January for UVOT. The first GRB was detected by BAT on the 17th of December 2004, while the first X–ray afterglow was detected by XRT on the 23rd of the same month. The first optical afterglow was detected by UVOT on the 15th of March 2005. Since then and through September 2006 BAT detected 194 GRBs. The verification and calibration phase has ended on the 5th of April 2005 and since then all the data from the 3 instruments are publicly available together with the reduction and analysis software. In the period April 2005-September 2006 BAT detected 151 bursts. The average detection rate is 1 in 3.7 days, i.e. 2 each week. In the 80% of the cases the narrow field instruments satrted the observation before 300 seconds with aa average of 97±50 seconds. In the remaining cases the beginning of the observation was delayed by approximately 60 min due to the Earth occultation constraints. XRT detected 137 X–ray afterglows, UVOT detected 38 optical transients. Taking into account the optical transient detected by on–ground telescope 68 optical/IR afterglows have been detected out of 151 detected bursts and

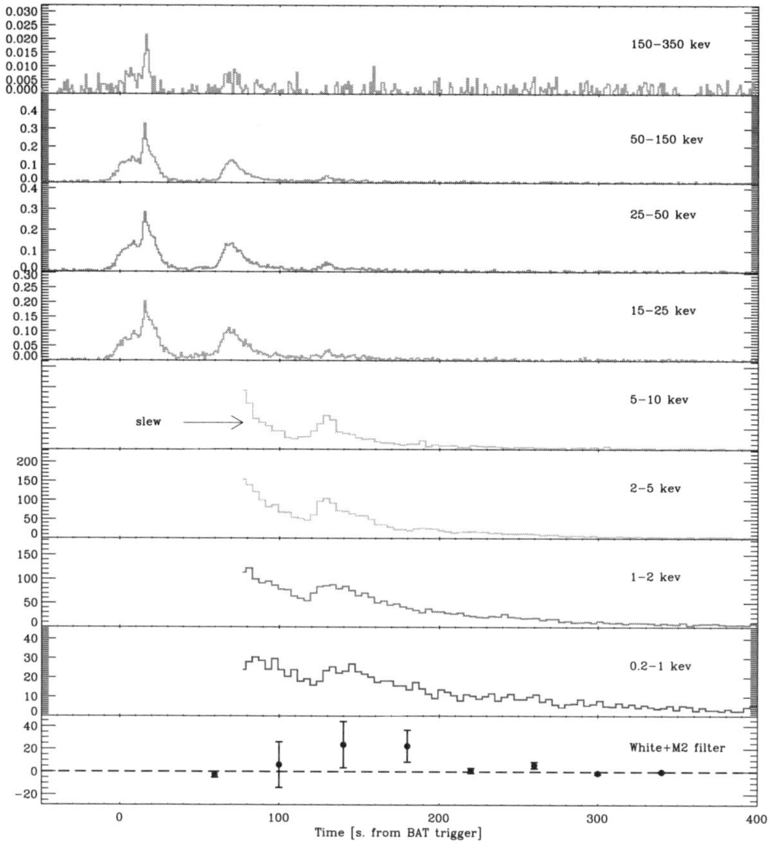

Fig. 4. The burst end early afterglow light curve of the GRB 060814 of the 3 *Swift* instruments in different energy bands .

for 42 of them the redshift has been measured (Fig. 3).

3. The X–ray afterglows

One of the most important scientific/observational result achived by *Swift* is the observation of the very early X–ray afterglow on a very large sample of GRB (> 160). These observations are revealing valuable and unprecedented information about GRBs, including the prompt emission–afterglow transition, emission site, central engine activity, forward-reverse shock physics, and the GRB immediate environment. Before *Swift*, X-ray afterglow emission was detected, in most cases, only several hours after the burst, by

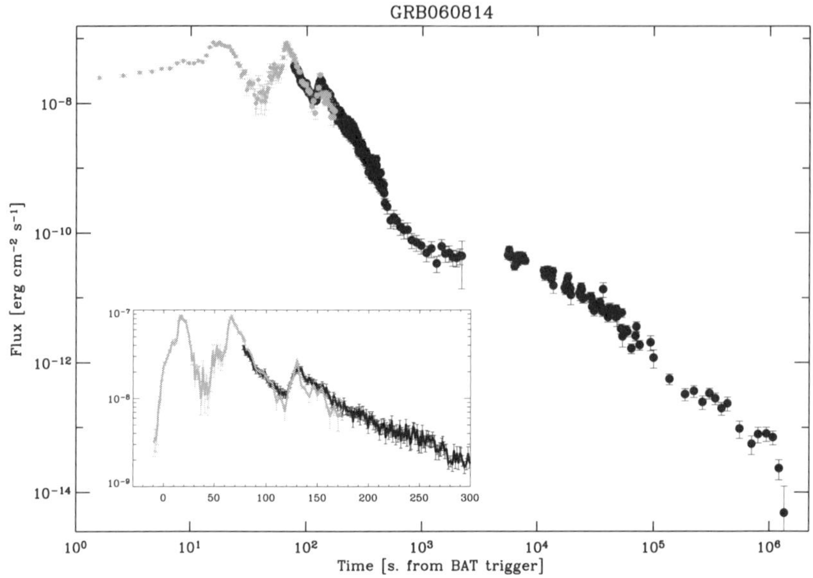

Fig. 5. The X–ray light curve of GRB 060814 in the 0.3-10 keV band. Black points are from XRT data, while grey points are the BAT data extrapolated to 0.3-10 keV band

which time the flux typically showed a smooth single power law decay $\sim t^{-1}$. However, the early afterglow evolution, in the first few hours, remained largely unexplored. Systematic analysis of the light curves on statistically significant *Swift* samples have revealed several common features (e.g. [6], [7], [8]). During the first few hundred seconds, a steep decay is often observed $\alpha \approx 3$–5, usually interpreted as the tail emission from the prompt GRB. This phase is followed by a much flatter decline, $\alpha \approx 0$–0.7, lasting up to 10^3–10^5 s (and in some cases even longer). Then the light curve steepens again, leading to $\alpha \approx 1$–1.5; it should be noted that this is consistent with those seen in previous missions since they typically started observations hours after the burst. At late times, a further steepening is sometimes observed, likely the signature of a jetted outflow. In $\sim 30\%$ of the afterglows bumps and flares appear superimposed to the power-law decay, for up to several tens of ks after the prompt GRB emission.

As an example of the data that *Swift* is collecting we consider the case of the observation of GRB 060814 ([9]). In the prompt light curve observed by BAT there are four main peaks (Fig.4), with a total fluence in the 15-150 keV band is 1.50±0.02 10^{-5} erg cm^{-2}. The time-averaged spectrum is

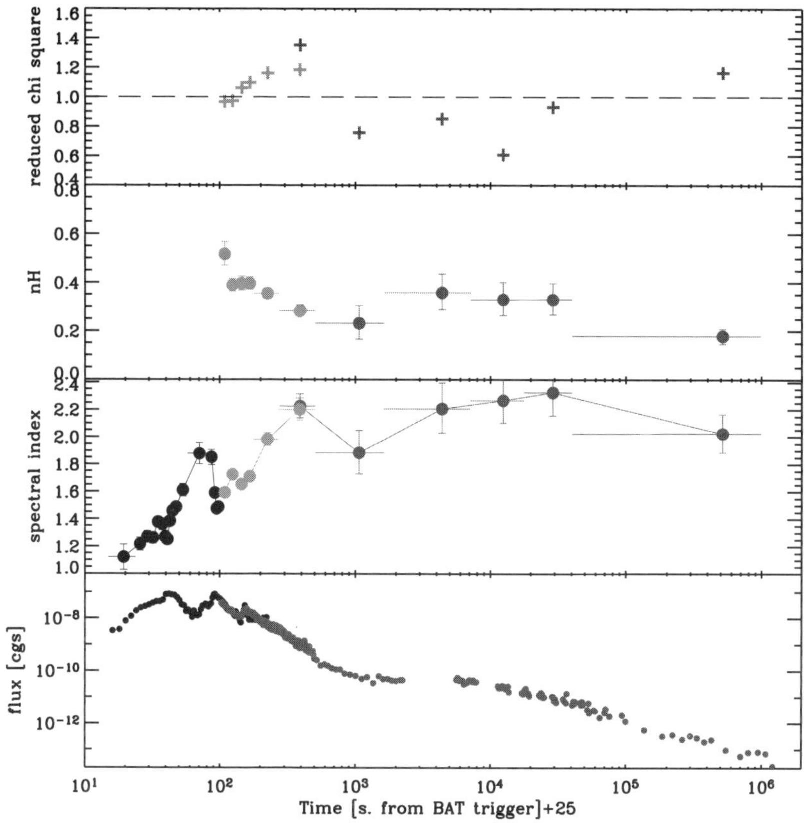

Fig. 6. Spectral evolution in the emission of the prompt and afterglow of GRB060814; black points are from BAT data only; grey points are from the combiantion of simulataneous BAT and XRT data; light grey points are from XRT data only. The origin of the time axis has been shifted by 20 seconds for the clarity of the logaritmic plot. In the first three panels (from the top) the reduced χ^2, the absorbing column and the photon index are respectively shown. In the bottom panel the light curve is shown.

best fit by a simple power-law model with a photon index of 1.58$\pm$0.02. The NFI instruments began observing the field 72 seconds after the BAT trigger. The XRT light-curve, initially, shows a steep decay (α_1=2.3) up to a break at t$\sim$1300s. On top of this steep decay there is a flare peaking at 150 s, coincident with the fouth peak of the prompt event seen by BAT. After the break, the afterglow shows a shallow decay (α_2=2.3) up to t=7500, when the light-curve steepens (α_2=2.3) (Fig.5). The prompt event was observed also by Konus-Wind ([10]) and SUZAKU ([11]). They both found that the best fit to the prompt emission spectrum is given by a power law with exponential cutoff model with photon index of 1.43±0.15 energy peak= 257^{+122}_{-58} keV and α=1.6±0.5, energy peak of 312^{+159}_{-152} keV respectively. These data are consistent between each other and are fully consistent with the spectrum observed by BAT. The brightness of the burst allowed us the study of the spectral evolution. Through the whole event from the prompt up to the late afterglow, *Swift* data can be best fitted with single power law model with highly variable photon index. In particular, the photon index has a minimum (i.e. the spectrum becomes harder) at the beginning of the prompt emission and softens as long as the next flare starts, when it abrubtly becomes harder (but a little bit softer than the previous); during all the prompt event the overall trend is a softening with the photon index increasing by about 1, reaching a value consistent with the later afterglow spectrum by the end of the last peak (Fig.6).

The Swift mission life was predicted to end, pre-launch, with orbital decay in about 2012, but new post-launch predictions now estimate orbital reentry no earlier than 2021. Although the design life of the spacecraft is based on a minimum five year mission, there are no known consumables or life-limiting design failure components that would limit the on-orbit lifetime.

References

1. Gehrels N., et al., 2004, ApJ, 611, 1005
2. Barthelmy, S. D., 2005, SSRev 120, 143
3. Burrows, D. N. et al., 2005, SSRev 120,165
4. Roming, P. W. A. et al., 2005, SSRev 120, 95
5. Nousek, J.A., 2006, proceedings of *Swift and GRBs: Unveiling the Relativistic Universe*, Venice, Italy, 5-9 June 2006
6. Tagliaferri, G., et al., 2005, Nature, 436, 985
7. Nousek, J. A., et al., 2005 ApJ
8. Burrows D.,N., et al., 2005, Science, 309,1833
9. Moretti, A., et al., 2006, GCN Circular 5447
10. Golenetskii S., et al., 2006, GCN Circular 5460
11. Suzuki, M., et al., 2006, GCN Circular 5487

GAMMA-RAY ASTROPHYSICS WITH AGILE

F. LONGO*, M. TAVANI, G. BARBIELLINI, A. ARGAN, M. BASSET, F. BOFFELLI, A. BULGARELLI, P. CARAVEO, A. CHEN, E. COSTA, G. DE PARIS, E. DEL MONTE, G. DI COCCO, I. DONNARUMMA, M. FEROCI, M. FIORINI, L. FOGGETTA, T. FROSLYAND, M. FRUTTI, F. FUSCHINO, M. GALLI, F. GIANOTTI, A. GIULIANI, C. LABANTI, I. LAPSHOV, F. LAZZAROTTO, F. LIELLO, P. LIPARI, M. MARISALDI, M. MASTROPIETRO, E. MATTAINI, F. MAURI, S. MEREGHETTI, E. MORELLI, A. MORSELLI, L. PACCIANI, A. PELLIZZONI, F. PEROTTI, P. PICOZZA, C. PITTORI, C. PONTONI, G. PORROVECCHIO, M. PREST, G. PUCELLA, M. RAPISARDA, E. ROSSI, A. RUBINI, P. SOFFITTA, A. TRACI, M. TRIFOGLIO, A. TROIS, E. VALLAZZA, S. VERCELLONE, A. ZAMBRA, D. ZANELLO

The AGILE collaboration, `http://agile.iasf-roma.inaf.it`

The AGILE Mission will explore the gamma-ray Universe with a very innovative instrument combining for the first time a gamma-ray imager (sensitive in the range 30 MeV - 50 GeV) and a hard X-ray imager (sensitive in the range 15-45 keV). AGILE will be operational at the beginning of 2007 and it will provide crucial data for the study of Active Galactic Nuclei, Gamma-Ray Bursts, unidentified gamma-ray sources, Galactic compact objects, supernova remnants, TeV sources, and fundamental physics by microsecond timing.

Keywords: Space instrumentation, gamma-rays, X-rays

1. INTRODUCTION

The space program AGILE (*Astro-rivelatore Gamma a Immagini LEggero*) is a high-energy astrophysics Mission supported by the Italian Space Agency (ASI) with scientific and programmatic participation by INAF, INFN and several Italian universities [5]. The industrial team includes Carlo Gavazzi Space, Alcatel-Alenia-Space-Laben, Oerlikon-Contraves, and Telespazio.

The main scientific goal of the AGILE program is to provide a powerful and cost-effective mission with excellent imaging capability simultaneously

*corresponding author: e-mail: `francesco.longo@ts.infn.it`

Fig. 1. *Left panel*: The integrated AGILE satellite in its final configuration. The total satellite mass is equal to 350 kg. *Right panel*: the AGILE scientific instrument showing the hard X-ray imager, the gamma-ray Tracker, and Calorimeter. The Anticoincidence system is partially displayed, and no lateral electronic boards and harness are shown for simplicity. The AGILE instrument "core" is approximately a cube of about 60 cm size and of weight equal to 100 kg.

in the 30 MeV-50 GeV and 15-45 keV energy ranges with a very large field of view [6,7]. AGILE is currently (October 2006) carrying out the satellite qualification tests after having completed the gamma-ray calibration in November 2005. The Indian rocket PSLV is planned to launch AGILE during the spring of 2007.

This paper is essentially based on the last review written by the AGILE collaboration on the AGILE instrument and its science objectives [8].

The instrument is very compact (see Fig. 1) and light ($\sim$ 120 kg) and it is aimed at detecting new transients and monitoring gamma-ray sources within a very large field of view (FOV $\sim 1/5$ of the whole sky). The total satellite mass is equal to 350 kg.

AGILE is expected to substantially advance our knowledge in several research areas including the study of Active Galactic Nuclei and massive black holes, Gamma-Ray Bursts (GRBs), the unidentified gamma-ray sources, Galactic transient and steady compact objects, isolated and binary pulsars, pulsar wind nebulae (PWNae), supernova remnants, TeV sources, and the Galactic Center. Furthermore, the fast AGILE electronic readout and data processing (resulting in detectors' deadtimes smaller than $\sim 200\ \mu$sec) allow for the first time a systematic search for sub-millisecond gamma-ray/hard X-ray transients that are of interest for both Galactic objects (searching outburst durations comparable with the dynamical timescale of $\sim 1\,M_{\odot}$ compact objects) and quantum gravity studies [11] .

The AGILE Science Program will be focused on a prompt response to gamma-ray transients and alert for follow-up multiwavelength observations. AGILE will provide crucial information complementary to several space missions (Chandra, INTEGRAL, XMM-Newton, SWIFT, Suzaku) and it will support ground-based investigations in the radio, optical, and TeV bands. Part of the AGILE Science Program will be open for Guest Investigations on a competitive basis. Quicklook data analysis and fast communication of new transients will be implemented as an essential part of the AGILE Science Program.

2. The Scientific Instrument

The AGILE scientific payload is made of three detectors combined into one integrated instrument with broad-band detection and imaging capabilities [1,6]. The Anticoincidence and Data Handling systems complete the instrument. Table 1 summarizes the instrument scientific performance.

The Gamma-Ray Imaging Detector (GRID) is sensitive in the energy range $\sim$ 30 MeV–50 GeV, and consists of a Silicon-Tungsten Tracker, a Cesium Iodide Calorimeter, and the Anticoincidence system. It is characterized by a very fine spatial resolution (obtained by a special arrangement of Silicon microstrip detectors and analog signal storage and processing) and by the smallest ever obtained deadtime for gamma-ray detection (200 μs). The GRID is designed to achieve an optimal angular resolution (source location accuracy $\sim 15'$ for intense sources), an unprecedentedly large field-of-view ($\sim$ 2.5 sr), and a sensitivity comparable to that of EGRET for sources within 10-20 degree off-axis (and substantially better for larger off-axis angle for the same exposure).

The hard X-ray imager (Super-AGILE) is the unique characteristic of the AGILE instrument. This imager is placed on top of the gamma-ray detector and is sensitive in the 15-45 keV band. It has an optimal angular resolution (6 arcmin) and a good sensitivity over a $\sim$ 1 sr field of view. The main characteristic of AGILE will be then the possibility of simultaneous gamma-ray and hard X-ray source detection with arcminute positioning and on-board GRB/transient source alert capability.

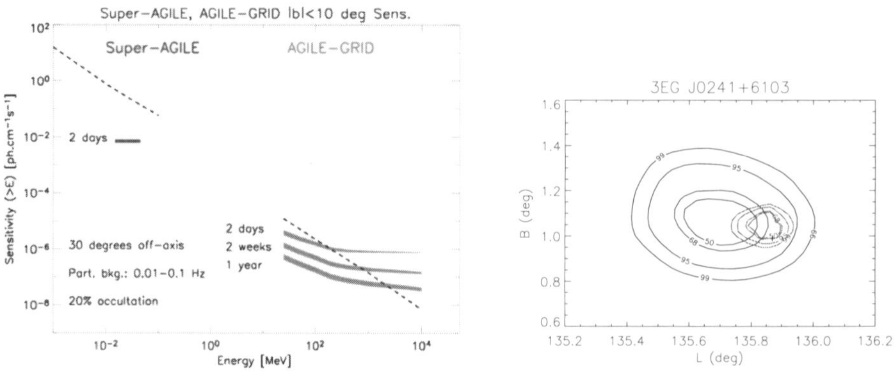

Fig. 2. *Left panel*: simulated integrated flux sensitivity of AGILE (GRID and SA) as a function of energy for a 30-degree off-axis source in the Galactic plane. The Crab spectrum is shown by the dotted line. *Right panel*: simulated positioning of a gamma-ray source in the Galactic plane obtained by the AGILE Likelihood and diffuse gamma-ray background processing [3]. The (more compact) red contours show the 50, 68, 95, 99 percent confidence levels obtained for the AGILE-GRID; the black contours show the confidence levels obtained with the EGRET likelihood analysis.

A Mini-Calorimeter operating in the "burst mode" is the third AGILE detector. It is part of the GRID, but is also capable of independently detecting GRBs and other transients in the 400 keV - 100 MeV energy range with optimal timing capabilities.

Fig. 1 shows the integrated AGILE satellite and a representation of the instrument. More detailed information is presented elsewhere [8,10].

3. Science with AGILE

The AGILE instrument has been designed and developed to obtain an excellent imaging capability in the energy range 100 MeV-50 GeV, improving the EGRET angular resolution by a factor of 2, see Fig. 2; AGILE will have a very large field-of-view for both the gamma-ray imager (2.5 sr) and the hard X-ray imager (1 sr) (FOV larger by a factor of ~6 than that of EGRET); The excellent timing capabilities of AGILE, with overall photon absolute time tagging of uncertainty below $2\,\mu$s and very small deadtimes ($200\,\mu$s for the GRID, $\sim 5\,\mu$s for the sum of the SA readout units, and $\sim 20\,\mu$s for each of the individual CsI bars) allows fundamental physics studies; AGILE will achieve a good sensitivity for point sources, comparable to that of EGRET in the gamma-ray range for sources within 20 degrees off-axis (except a central region of smaller effective area), and very flat up to 50-60 degrees off-axis. Depending on exposure and the diffuse background , the flux sensitivity threshold can reach values of $(10-20)\times 10^{-8}$ ph.cm^{-2} s^{-1} above 100 MeV. The hard X-ray imager sensitivity is between 10 and 20 mCrab at 20 keV for a 1-day integration over a field of view near 1 sr; Furthermore AGILE will have a good sensitivity to photons in the energy range ~30-100 MeV, with an effective area above 200 cm^2 at 30 MeV; and a very rapid response to gamma-ray transients and gamma-ray bursts, obtained by a special quicklook analysis program and coordinated ground-based and space observations. AGILE will provide accurate localization (~2-3 arcmins) of GRBs and other transient events obtained by the GRID-SA combination (for typical hard X-ray transient fluxes above ~1 Crab); the expected GRB detection rate is $\sim 1-2$ per month; and long-timescale continuous monitoring (~2-3 weeks) of gamma-ray and hard X-ray sources; Finally the AGILE satellite can be repointed after special alerts (~1 day) in order to position the source within the Super-AGILE FOV to obtain hard X-ray data for gamma-ray transients detected by the GRID in the external part of its FOV or by other high-energy missions (e.g., GLAST).

The combination of simultaneous hard X-ray and gamma-ray data will provide a formidable combination for the study of high-energy sources.

Figs. 2 shows and example of the expected positioning of a gamma-ray source in the Galactic plane. The GRID configuration will achieve a PSF with 68% containment radius better within $1^\circ - 2^\circ$ at E $>$ 300 MeV allowing a gamma-ray source positioning with error box radius near $10^{'}-20^{'}$ depending on source spectrum, intensity, and sky position. Super-AGILE operating in the 15-45 keV band has a spatial resolution of 6 arcminute (pixel size). This translates into a positional accuracy of 1-3 arcmins for relatively strong transients at the Crab flux level.

A crucial feature of AGILE is its large field of view for both the gamma-ray and hard X-ray detectors. Fig. 3 shows typical gamma-ray FOVs obtained for a sequence of AGILE pointings. Relatively bright AGNs and Galactic sources flaring in the gamma-ray energy range above a flux of 10^{-6} ph cm^{-2} s^{-1} can be detected within a few days by the AGILE Quick-look Analysis. We conservatively estimate that for a 2-year mission AGILE is potentially able to detect hundreds of gamma-ray flaring AGNs and other transients. The very large FOV will also favor the detection of GRBs above 30 MeV. Taking into account the high-energy distribution of GRB emission above 30 MeV, we conservatively estimate that ~1 GRB/month can be detected and imaged in the gamma-ray range by the GRID. Super-AGILE will be able to detect about 30% of the sources detected by INTEGRAL [2]; about 10 hard X-ray sources per day are expected to be detected by SA for typical pointings of the Galactic plane.

The existence of a large number of variable gamma-ray sources (extragalactic and near the Galactic plane [4]) makes necessary a reliable program for quick response to transient gamma-ray emission. Quicklook Analysis of gamma-ray data is a crucial task to be carried out by the AGILE Team. Prompt communication of gamma-ray transients (that require typically 2-3 days to be detected with high confidence for sources above $10^{-6}\,\mathrm{ph\,cm^{-2}\,s^{-1}}$) will be ensured. Detection of short timescale (seconds/minutes/hours) transients (GRBs, SGRs, and other bursting events) is possible in the gamma-ray range. A primary responsibility of the AGILE Team will be to provide positioning of short-timescale transient as accurate as possible, and to alert the community though dedicated channels.

The AGILE average exposure per source will be larger by a factor of ~ 4 for a 1-year sky-survey program compared to the typical exposure obtainable by EGRET for the same time period. Deep exposures for selected regions of the sky can be obtained by a proper program with repeated overlapping pointings. This can be particularly useful to study selected Galactic and extragalactic sources.

AGILE detectors have optimal timing capabilities. The on-board GPS system allows to reach an absolute time tagging precision for individual photons better than $2\,\mu$s. Depending on the event characteristics, absolute time tagging can achieve values near $1 - 2\,\mu$s for the Silicon-Tracker, and $3 - 4\,\mu$s for the Mini-Calorimeter and Super-AGILE. Instrumental deadtimes will be unprecedentedly small for gamma-ray detection. The GRID deadtime will be lower than $200\,\mu$s (improving by almost three orders of magnitude the performance of previous spark-chamber detectors such as EGRET). Taking into account the segmentation of the electronic readout of MCAL and Super-AGILE detectors (30 MCAL elements and 16 Super-AGILE elements) the MCAL and SA effective deadtimes will be less than those for individual units. We obtain $\sim 2\,\mu$s for MCAL, and $5\,\mu$s for SA. Furthermore, a special memory will ensure that MCAL events detected during the Si-Tracker readout deadtime will be automatically stored in the GRID event. For these events, precise timing and detection in the $\sim$ 1–200 MeV range can be achieved with temporal resolution well below $100\,\mu$s. This may be crucial for AGILE high-precision timing investigations.

4. Multiwavelength Observations Program

The scientific impact of a high-energy Mission such as AGILE (broad-band energy coverage, very large fields of view) is greatly increased if an efficient

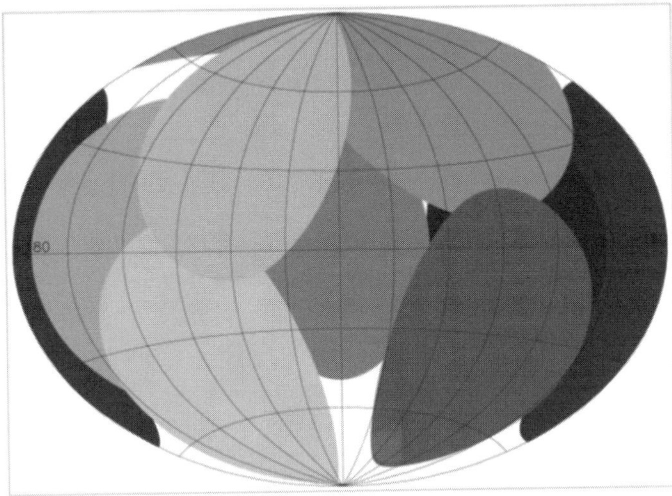

Fig. 3. An example of a sequence of AGILE pointings showing the very large gamma-ray field of view (in Galactic coordinates). Typical AGILE pointings will last 3 weeks and are subject to solar panel constraints.

program for fast follow-up and/or monitoring observations by ground-based and space instruments is carried out. The AGILE Science Program will overlap and be complementary to those of many other high-energy space Missions (INTEGRAL, XMM-Newton, Chandra, SWIFT, Suzaku, GLAST) and ground-based instrumentation.

The AGILE Science Program will involve a large astronomy and astrophysics community and emphasizes a quick reaction to transients and a rapid communication of crucial data. Past experience shows that in many occasions there was no fast reaction to γ-ray transients (within a few hours/days for unidentified gamma-ray sources) that could not be identified. AGILE will take advantage, in a crucial way, of the combination of its gamma-ray and hard X-ray imagers.

The AGILE Science Group (ASG) (established in 2001) aims at favoring the scientific collaboration between the AGILE Team and the community especially for coordinating multiwavelength observations based on AGILE detections and alerts. The ASG is open to the international astrophysics community and consists of the AGILE Team and qualified researchers contributing with their data and expertise in optimizing the scientific return of the Mission. Several working groups are operational on a variety of scientific topics including blazars, GRBs, pulsars, and Galactic compact objects.

The AGILE Team is also open to collaborations with individual observing groups.

Updated documentation on the AGILE Mission can be found at the web site http://agile.iasf-roma.inaf.it.

References

1. Barbiellini G. et al., 2000, Proceedings of the 5th Compton Symposium, AIP Conf. Proceedings, ed. M. McConnell, Vol. 510, p. 750
2. Bird, A.J., et al., 2006, ApJ, 636, 765
3. Chen, A., et al., 2006, in preparation
4. Tavani, M., et al., 1997, ApJ, 479, L109.
5. Tavani M., Barbiellini G., Caraveo P., Di Pippo S., Longo F., Mereghetti S., Morselli A., Pellizzoni A., Picozza P., Severoni S., Tavecchio F., Vercellone S., 1998, *AGILE Phase A Report*
6. Tavani M., et al., 2000, Proceedings of the 5th Compton Symposium, AIP Conf. Proceedings, ed. M. McConnell, Vol. 510, p. 746
7. Tavani M., 2003, Texas in Tuscany, XXI Symposium on Relativistic Astrophysics, Florence, Italy, 9-13 December 2002, p. 183
8. Tavani M., et al., 2006, paper presented at the SPIE Conference n. 6266, Orlando, May 24, 2006; to be published in the Conf. Proceedings.
9. Tavani, M., et al., 2006, "The AGILE Scientific Instrument", to be submitted to NIM.
10. Tavani, M., et al., 2006, "The AGILE Mission and its Scientific Program", in preparation.
11. Tavani, M., 2006, in preparation.

The GLAST Mission

J. E. McEnery* for the GLAST Mission team

NASA/GSFC, Code 661
Greenbelt, MD 20740, USA
**E-mail: mcenery@milkyway.gsfc.nasa.gov*

The Gamma-ray Large Area Space Telescope (GLAST), scheduled for launch in late 2007, is a satellite based observatory to study the high energy gamma-ray sky. There are two instruments on GLAST: the Large Area Telescope (LAT) which provides coverage from 20 MeV to over 300 GeV, and the GLAST Burst Monitor (GBM) which provides observations of transients from 8 keV to 30 MeV. GLAST will provide capabilities well beyond those achieved by the highly successful EGRET instrument on the Compton Gamma-Ray Observatory, with dramatic improvements in sensitivity, angular resolution and energy range. The very large field of view will make it possible to observe ~20% of the sky at any instant, and the entire sky on timescales of a few hours.

1. Introduction

During the 1990's the field of high-energy gamma-ray astrophysics blossomed. The EGRET experiment on the Compton Gamma-Ray Observatory (CGRO) revolutionized GeV astrophysics with the detection of several hundred new gamma-ray sources.[1] EGRET greatly advanced our understanding of the high energy gamma-ray sky: Among them were the discoveries that for many blazars, intense gamma-ray flares are seen, that many blazars and some pulsars have peak luminosity at GeV energies, and that the spectrum of GRBs extends to at least GeV energies. However, of this multitude of sources, less than half have been definitively associated with known astrophysical objects. Thus, most of the gamma-ray sky, as we currently understand it, consists of unidentified objects. EGRETs great discoveries have left us with many intriguing puzzles for GLAST, the next generation GeV gamma-ray instrument to uncover.

The main instrument on GLAST is the Large Area Telescope (LAT), which will survey the sky from 20 MeV to over 300 GeV. The LAT provides greatly superior performance compared with previous missions in this energy band; it has more than an order of magnitide greater sensitivity; it has a much broader energy range; it has a very wide field of view which allows survey mode observations where the entire sky is observed every 3 hours. The GLAST LAT will open several new windows of discovery space; ***spectrally***, the LAT will open a new window in the range betwen 10 GeV and 100 GeV; ***temporally*** the LAT will provide unbiased monitoring of all gamma-ray sources in the sky down to timescales of hours and ***spatially*** the improved angular resolution will allow spatially resolved studies of extended GeV sources.

A second intrument on GLAST, the Glast Burst Monitor (GBM) provides complementary observations of gamma-ray bursts from 8 keV to 30 MeV. The GBM alone will provide excellent prompt spectral observations of hundreds of bursts per year. The GBM plays a crucial role in connecting the ground-breaking high-energy measurments made by the LAT to the better studied low-energy regime. The GBM will provide real-time burst locations over a wide field of view with sufficient accuracy to allow repointing the GLAST spacecraft. Time-resolved spectra of many bursts recorded by the LAT and the GBM will allow the investigation of the relation between the keV and MeV-GeV emission from GRBs over an unprecedented seven decades of energy.

2. GLAST Instruments

2.1. *Large Area Telescope*

The primary instrument is the GLAST Large Area Telescope (LAT), which is sensitive at energies from 20 MeV to 300 GeV. Like EGRET, the LAT will detect gamma-rays through conversion to electron positron pairs and measurement of their direction in a tracker and their energy in a calorimeter. It uses a segmented plastic scintillator anti-coincidence system to provide rejection of the intense background of charged cosmic-rays. A schematic of the LAT is shown in Figure 1.

Along with increased area with respect to EGRET, the design of each element has been refined to improve the sensitivity and angular resolution and to extend the energy range to higher energy.

In the tracker,[2] incident photons convert to electron-positron pairs in one of 16 layers of lead converter and are tracked by single-sided solid-state silicon strip detectors (SSD) through successive planes (the strips are centered every 228 microns). Tracking the pairs allows the reconstruction of the incident photon's arrival direction.

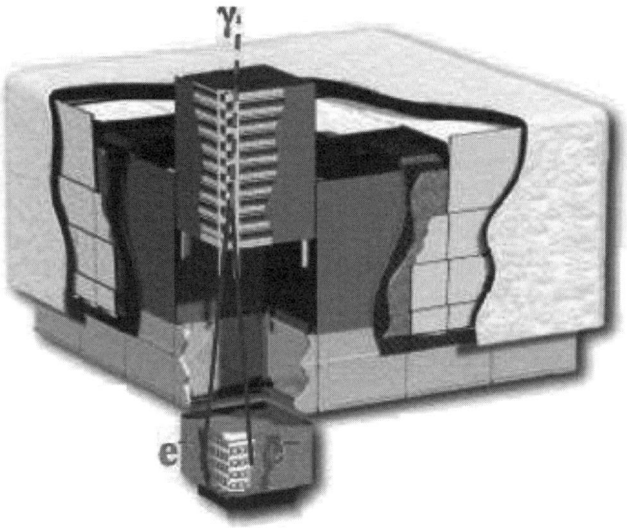

Fig. 1. The GLAST Large Area Telescope

There are a total of 36 SSD planes, 18 x,y pairs. The first 12 pairs (the front section of the tracker) are covered by tungsten plates that are each 0.03 radiation lengths (r.l.) thick, the next 4 are covered by converters 0.18 r.l. thick, and the final two planes have no converter. The photons that convert in the front section (0.36 r.l.) give the highest angular resolution, while the back section assures that the overall photon detection efficiency is as high as practical. The last two layers make sure we know the entry point of the particles into the calorimeter, where the tracking in the crystal layers, described below, continues.

The calorimeter consists of a segmented arrangement of 1536 CsI (Tl) crystals in 8 layers, giving both longitudinal and transverse information about the energy deposition pattern. This has several advantages over the shallower, monolithic calorimeter used in EGRET. The segmentation allows for the shower development in the calorimeter to be reconstructed resulting in a method of correction for energy leaking out the bottom of the calorimeter. Combined with the greater depth of the calorimeter this provides much the required energy resolution at high energies up to several hundred GeV. In addition, the segmentation allows the calorimeter to provide direction information of the gamma-ray which can be used if the gamma-ray did not convert in the tracker and provide important pattern recognition inputs for the rejection of background.

The AntiCoincidence Detector[3] (ACD) provides most of the rejection of charged particle backgrounds. It consists of an array of 89 plastic scintillator tiles each with charged particle detiction efficiency of 0.9997 read out by wave-shifting fibers and photomultiplier tubes. The segmentation prevents loss of effective area at high energy due to self veto caused by low energy photons coming from the cascade showers in the calorimeter. This was an issue in EGRET that occured when albedo from showers in their calorimeter fired the monolithic ACD causing a 50% reduction in effective area at 10 GeV. The ACD has been designed so the LAT will have at least 80% efficiency for 300 GeV photons.

2.2. *GLAST Burst Monitor*

The GLAST Burst Monitor is designed to provide observations of gamma-ray bursts from 8 keV to 30 MeV. The GBM consists of 12 sodium iodide (NaI) detectors, which provide GRB triggering, localisation and spectra from 8 keV to 1 MeV, and 2 Bismuth Germanate (BGO) detectors which provide spectral measurments from 150 keV to 30 MeV, thus overlapping with the LAT energy range.

The GBM can detect gamma-ray bursts over the entire unocculted sky (9.5 sr) and will peform several crucial functions. It will provide a GBM burst alert which is transmitted to the LAT and to the ground, it will determine the position of the burst which will allow autonomous repointing of the spacecraft to keep the GRB within the LAT FoV for high energy afterglow studies. Finally (and most importantly) the GBM will provide low-energy context measurement of the burst lightcurve and energy spectrum.

3. GLAST Science Operations

The GLAST observatory will be flown in a low-Earth orbit (altitude of 565 km and inclination of 25.3 degrees) and can operate in both a "survey" mode and a "stare" mode.

The primary goal of survey mode is to provide the most uniform sky coverage with the LAT over a short time. Survey mode maximizes the ability to study the behaviour of time variable sources anywhere in the sky, over a wide range of timescales. The current concept for survey mode is to slew once per orbit, alternating between 35 degrees above and 35 degrees below the orbit plane, providing gamma-ray observations of the entire sky every two orbits (190 mins)

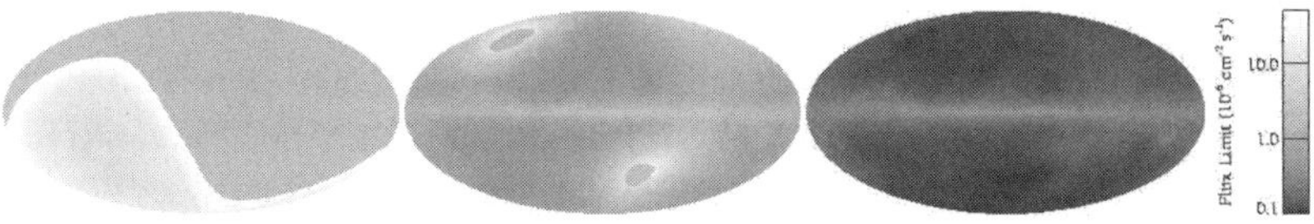

Fig. 2. LAT sensitivity for three observation durations: 100 s, an orbit and a day

The large LAT field of view and the uniformity of sky coverage in survey mode is illustrated in Figure 2, which shows the the LAT sensitivity for instantaneous (100 second), one orbit and one day observations.

The science data collected by the two instruments on GLAST is stored onboard on a solid state recorder and from there is transfered to the ground six to eight times per day. The raw data is processed at the Mission Operations Center at GSFC and then the LAT data is sent to the LAT Instrument Science Operations Center (ISOC) at SLAC and the GBM data is sent to the GBM instrument operations center at MSFC.

The ISOC will produce and deliver processed LAT data to the GLAST Science Support Center for distribution to the scientific community. Automated Science Processing (ASP) will be performed as soon as the LAT data are available (typically within 3-6 hours), for time critical science analyses: GRB detection and follow up, and flaring source monitoring and detection. The processed LAT and GBM data will be made available to the scientific community through the GLAST Science Support Center at GSFC.

Both the LAT and the GBM perform onboard searches from GRB. Both instruments will send an alert to the ground (GCN) within 10 - 15 seconds. These alerts will contain information about the location, flux and spectrum of the GRB. Either instrument can also initiate an autonomous repoint request to the spacecraft, which will then slew to keep the candidate GRB location within the LAT FOV for a selectable dwell time (initially five hours).

4. GLAST Science Performance

Figure 3 shows the LAT effective area and angular resolution for gamma-rays at normal incidence. The plots show three curves which correspond to different event selections, each tuned for different science goals. The middle curve is the default analysis, designed to have good background rejection, angular and energy resolution. The top curve corresponds to a looser event cuts, with poorer background rejection but higher effective area which will greatly benefit variability studies. The lower curve corresponds to events which converted in the front part of the tracker only and thus have excellent angular resolution, ideal for mapping spatially extended sources. We note that the optimisation of event selections is a very active area of study within the LAT collaboration and that significant improvements may be possible.

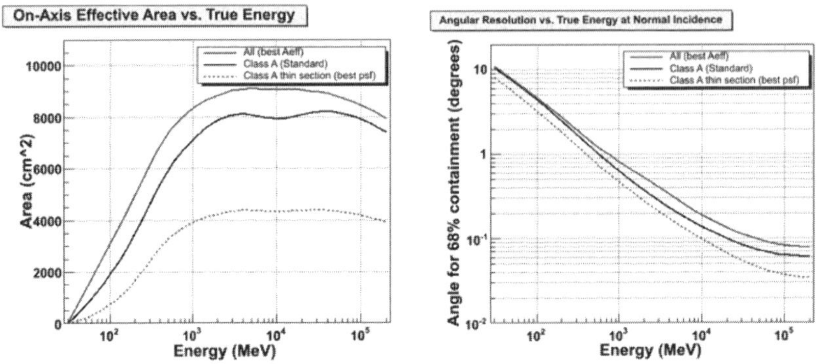

Fig. 3. LAT performance, showing the effective area and single photon angular resolution for gamma-rays at normal incidence. The performance from the standard event selection is described by the middle curve, also shown is the performance for event selections optimized for variability (top curves) and spatial (lower curves) studies.

Table 1. Properties of the LAT and GBM

	GBM	LAT
Energy Range	8 keV - 30 MeV	20 MeV - 300 GeV
Energy resolution	12% FWHM at 511 keV	10%
Field of View	9.5 sr	>2.0 sr
Angular Resolution	-	3.4° (100MeV)
		0.09° (10 GeV)
Onboard GRB locations	<15 deg within 1.8 s	-
Ground GRB locations	<5 deg within 30mins	-
Effective area	-	8000 cm^2
GRB sensitivity	0.71 ph cm-2 s-1	-
	(peak flux 50-300 keV)	-
Point source sensitivity	-	3×10^{-9}ph cm^{-2} s^{-1}

The increase in effective area will be of enormous importance for identifying sources via their temporal signatures. Flares from blazars will be detected at much lower flux levels and on far shorter time intervals. This greatly increases the likelihood of observing correlated variability in several wavebands, allowing a firm identification of the gamma-ray source. The extra collection area will also be very important for pulsar observations. The only Galactic sources of high energy gamma-rays that have been positively identified are all young rotation-powered pulsars, so it is likely that many of the unidentified sources are also pulsars. GLAST will be able to do blind periodicity searches on all EGRET unidentified sources and thus determine which are pulsars. The LAT will have the potential to detect gamma-ray variability from new classes of putative gamma-ray bright objects such as galactic microquasars or dim, decaying signals from gamma-ray burst afterglows.

The sky-survey capabilities of GLAST will revolutionize gamma-ray variability studies. The LAT will provide detailed unbiased sampling of the lightcurves of a large number of gamma-ray bright objects on a wide range of timescales. This is most easily illustrated graphically: in the left-hand panel of Figure 4 we show LAT observations of a 5-year simulated lightcurce (the average gamma-ray flux was the same as observed from PKS 0528 by EGRET).

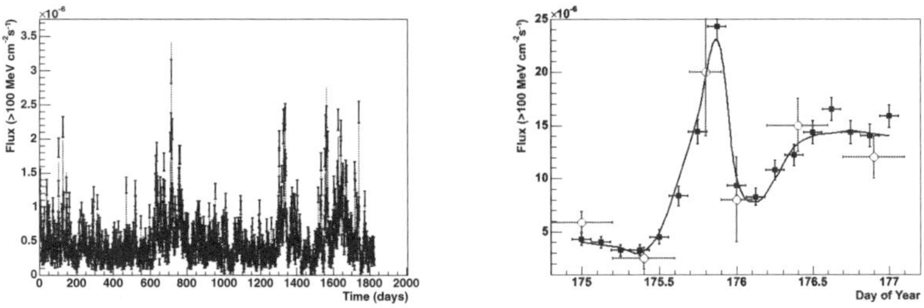

Fig. 4. (a) the left hand plot shows a 5 year LAT observation of a simulated lightcurve, (b) the right hand plot shows EGRET (open red circles) and simulated LAT (solid blue squares) observations of an intraday flare seen in 1995 from PKS 1622-297

The continuous sky monitoring means that the LAT will "catch" extraordinary flares. The large effective area will allow lightcurves to be resolved to much shorter timescales than has previously been possible at GeV energies. The right hand panel in Figure 4 shows EGRET observations of a rapid GeV flare from PKS 1622-297 in 1995,[4] a lightcurve consistent with the EGRET observations (black line), and simulated LAT survey mode observations. The LAT, in survey mode, would see a flare light this from any region in the sky at any time. This is extremely important as blazars are highly variable and it is anticipated that some objects will only be detectable during flares.

The theme of the meeting in Elba was on the study of variable and transient sources, and in discussing GLAST science performance we have focused on its capabilities to study these classes of object. One of the most dramatic classes of transient sources are gamma-ray bursts. GLAST will contribute to GRB studies by providing superb prompt spectra of GRB spanning seven decades in energy. The GBM will detect around 200 GRB/year of which 80 will lie within 65 degrees of the LAT boresight. GLAST observations of GRB will address many open questions about the high energy behaviour of bursts: is there a second high energy emission component? Does the MeV-GeV emission from GRB show different temporal behaviour to the routinely measured keV emission? How frequent are GeV afterglows and what is their nature? The answers to these questions will teach us much about the energetics and dynamics of GRB and lead us closer to a complete understanding of these enigmatic objects.

5. Guest Investigator Support and Data Availability

All public GLAST data, analysis software and documentation will be available at the GLAST science support center webpage.[5] Data releases in the first year of GLAST operations will be different from subsequent years. During year 1, all GBM data will be released, LAT summary data from a pre-determined list of approximately 20 source of interest will be provided. The summary data included time-binned spectra and associated errors and upper limits. Similarly, for any flaring source (flux $> 2 \times 10^{-6} cm^{-2} s^{-1}$, E$> 100 MeV$), summary information will be released. In addition, summary information for all GRB observations will be released. At the end of the first year, all the LAT photon candidate events will be released and subsequent event data will be released immediately. We note that after the first year, even for pointed observations, all data will be publicly available and there will be no proprietory GLAST data.

References

1. R. C. Hartman, et al *ApJ Supp.* **123**, 79 (1999).
2. W. B. Atwood, et al *submitted to Astroparticle Physics* (2007).
3. A. A. Moiseev, et al *Astroparticle Physics* **27**, 339 (2007).
4. J. R. Mattox, et al *ApJ* **476**, 692 (1997).
5. http://glast.gsfc.nasa.gov/ssc

II Ground-based Telescope

Recent Results from CANGAROO

MASAKI MORI* for the CANGAROO team
Institute for Cosmic Ray Research, University of Tokyo,
Kashiwa, 277-8582 Chiba, Japan
** E-mail: morim@icrr.u-tokyo.ac.jp*
http://icrhp9.icrr.u-tokyo.ac.jp/

The CANGAROO-III telescope system for very-high-energy gamma-ray astrophysics consists of four 10-m atmospheric Cherenkov telescopes located near Woomera, South Australia. The construction of the fourth telescope was completed in summer 2003, and stereoscopic observations have been in progress since March 2004. Here we report on the status of the system and some recent results from CANGAROO-III observations.

Keywords: gamma rays; observation; atmospheric Cherenkov telescopes

1. Introduction

CANGAROO is an acronym for the Collaboration of Australia and Nippon (Japan) for a GAmma Ray Observatory in the Outback. After successful operation of the 3.8m imaging Cherenkov telescope (CANGAROO-I) for 7 years, which was the first of this kind in the southern hemisphere, we constructed a new telescope of 7m diameter (CANGAROO-II) in 1999 next to the 3.8m telescope near Woomera, South Australia (136°47′E, 31°06′E, 160m a.s.l.). Then the construction of an array of four 10m telescopes (CANGAROO-III) was approved and as the first step the 7m telescope was upgraded to 10m diameter in 2000, with this becoming the first telescope of the CANGAROO-III array (T1) [1]. Results from observations with this first 10m telescope have been reported in publications (see, e.g. [1]).

In the following years, we have constructed an additional three 10m telescopes located at the corners of a diamond of 100m sides with improved mirrors, cameras and electronics. After tuning, we have started observation with the full system in stereo mode in March 2004. Here we report recent results from stereo observations with CANGAROO-III.

2. Stereo analysis: the case of the Crab nebula

The Crab nebula is the first established TeV gamma-ray source and is used as a calibration source to check performance of a Cherenkov telescope. However, from Woomera, it can be observed only at large zenith angles ($> 53°$). For stereo observations, the threshold energy of T1 is higher than other telescopes and thus we used the newer three telescopes (T2, T3 and T4 in the order of construction) for analysis. Because of the geometrical arrangement of the array, the effective baseline for large zenith angle observations becomes short which makes stereo reconstruction of images difficult.

To overcome the unfortunate situation described above, we developed new analysis methods [2]. To avoid the increased uncertainty of the intersection points, we introduced a new parameter, "IP distance" ($D_{\rm IP}$), which is defined as the distance between the intersection point and centroid of images. Then we searched for the intersection point which minimized the image widths and the difference between distance and $D_{\rm IP}$. This results in better angular resolution as seen in the θ^2 distribution in Monte Carlo simulations, where θ is the space angle between the source direction and the reconstructed arrival direction: gamma-ray signals should be seen as a peak toward $\theta^2 = 0$, whose sharpness depends on the angular resolution and the angular extent of a gamma-ray source.

We observed the Crab nebula in December 2003 in so-called "wobble" mode, changing the pointing directions $\pm 0.5°$ in declination from the target every 20 minutes, with a two-fold stereo mode (T2 and T3: T4 was not completed at that time). After basic data quality check, such as rejecting runs affected by clouds, a total of 890 minutes data were used for further analysis.

In addition to the conventional square cuts method using image parameters to enhance gamma-ray fractions, we applied two different analyses: the Likelihood method [3,4] and the Fisher Discriminant method [2,5]. In the latter method, effectiveness of the parameters for the gamma-ray-like event selection is evaluated using the simulation, and we can optimize the weights of the imaging parameters (*width* and *length* in this case) automatically by matrix inversion in estimating the probability of gamma-ray-like events, eliminating ambiguities in parameter cut positions. Finally we obtained the spectrum of the Crab nebula in the energy range from 2 to 20 TeV [5], which is consistent within the statistical and systematic errors with other measurements [6,7].

3. Recent results

3.1. *Pulsar PSR 1706-44*

A detection of a gamma-ray signal from PSR 1706-44, which was one of the EGRET-detected pulsars, was reported using the data acquired by CANGAROO-I 3.8m telescope [8]. The Durham group also reported a detection with their Mark 6 telescope [9]. H.E.S.S., however, claimed no detection from that direction [10]. We observed this source for 27 hours (ON) and 29 hours (OFF) with CANGAROO-III in May 2004. Preliminary analyses using T2 and T3 telescope pair did not show a peak in the θ^2 distribution [11]. The upper limit from this result is shown in Fig. 1, which is lower than the flux reported by CANGAROO-I. Further analysis is underway and the details will be reported elsewhere.

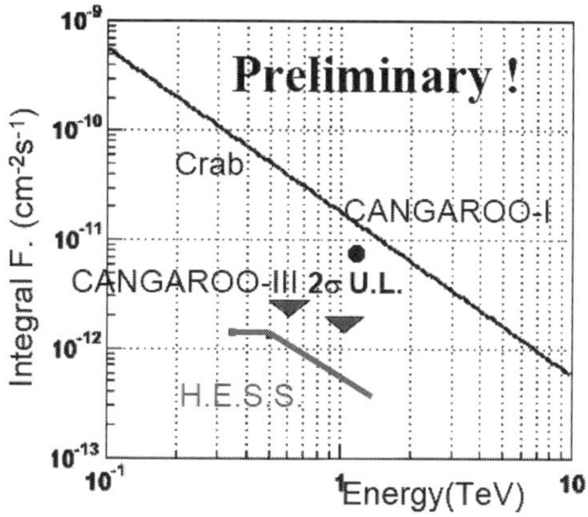

Fig. 1. Upper limits on gamma-ray flux from PSR 1706-44 from CANGAROO-III observations (triangle) [11]. The CANGAROO-I result is shown by a filled circle [8] and the H.E.S.S. limits [10] are also shown.

3.2. *Supernova remnant SN1006*

A detection of a gamma-ray signal from SN1006, which was shown to be a source of high-energy electrons through observation of non-thermal X-rays with ASCA [12], was reported using the data acquired by CANGAROO-I [13]. H.E.S.S., however, claimed no detection from that direction [14]. We observed this source for 27 hours (ON) and 29 hours (OFF) with CANGAROO-III in May 2004. Preliminary analyses using T2 and T3 telescope pair did not show any peaks in the θ^2 distribution for the NE-rim

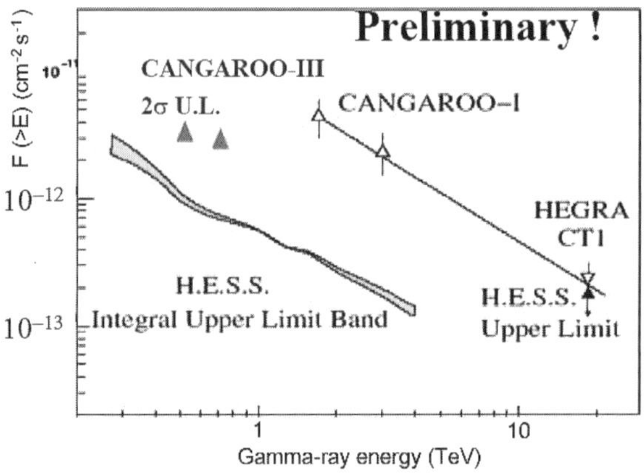

Fig. 2. Upper limits on gamma-ray flux from NE-rim of SN1006. The CANGAROO-I results are shown by open triangles [13] and the HEGRA CT1 result by inverted open triangle [15]. The CANGAROO-III upper limits [11] are shown by filled triangles with H.E.S.S. limits [19].

point which was the maximum point of the gamma-ray emission in the CANGAROO-I data [13]. The upper limit from this result is lower than the flux reported by CANGAROO-I. Further analysis is underway to check for possible extended emission, and the details will be reported elsewhere.

3.3. *Vela pulsar and nebula*

The Vela pulsar was observed in January/February 2004. After basic data quality check, a total of 1311 minutes data were used for further analysis [5], where the minimum elevation angle was set at 60°. The mean elevation angle was 70.9°, corresponding to an energy threshold of 600 GeV. The observations were carried out using the same wobble mode as for the Crab nebula observations. In this period, T2 and T3 were in operation, and we analyzed the stereo data from these two telescopes. For Vela, at a declination of −45°, the relative orientation of the two telescopes does not present any problems.

We used the optimized analysis procedure used for the Crab nebula analysis described above. The resulting θ^2 distribution for the Vela pulsar position showed no significant gamma-ray signal, giving upper limits as shown in Fig.3, which are consistent with H.E.S.S. results [19]. Also we did not see excess from the point offset by 0.13° from the pulsar, which was the maximum of the excess detected with the CANGAROO-I telescope [16].

The H.E.S.S. group detected a gamma-ray excess from the Vela X neb-

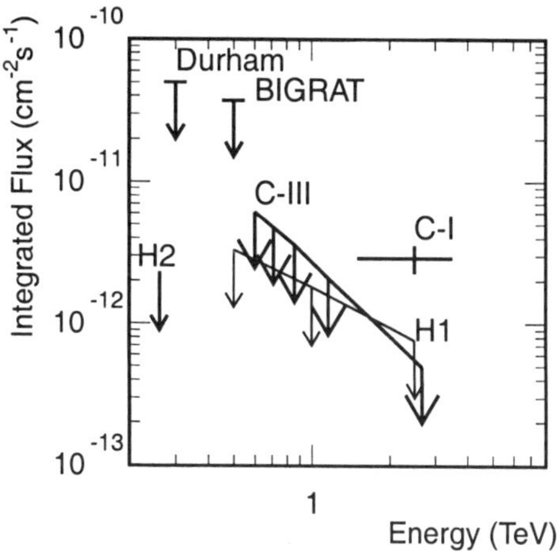

Fig. 3. The 2σ upper limits for the gamma-ray flux from the Vela pulsar by CANGAROO-III (C-III) [5]. C-I represents the CANGAROO-I excess from the point offset by 0.13° from the pulsar [16]. Also shown are upper limits reported by the Durham group [17], BIGRAT [18] and H.E.S.S. [19].

ula, extended over a 0.6° radius from the center of the emission [(R.A., decl.) $= (8^h35^m, -45°36')$, J2000] [19]. In order to analyze extended emission, we applied the following method. Gamma-ray-like events can be extracted by fitting position-by position F (Fischer discriminant) distributions under the assumption that gamma rays obey the Monte Carlo predictions, the proton background follows the average F distribution of all directions, and the total distribution is a linear combination of those two. We chose the background region to be more than 0.8° from the center, since we do not have sufficient statistics for off-source regions for these observations. The result of fitting is shown in Fig. 4. An excess was observed at $\theta^2 < 0.6$ deg^2 around the center of the Vela X region. The excess radius is marginally consistent with H.E.S.S. considering our angular resolution. The total number of gamma-ray-like events is 561 ± 114. Though the statistical significance is below the 5σ level, this could be supporting evidence of the H.E.S.S. detection. The differential fluxes for the excess regions are in general agreement with H.E.S.S. result [5].

3.4. *SNR RX J0852.0-4622*

We already reported a gamma-ray signal from this SNR using observations by CANGAROO-II in 2001 and 2002 [20]. This time we applied the Fisher

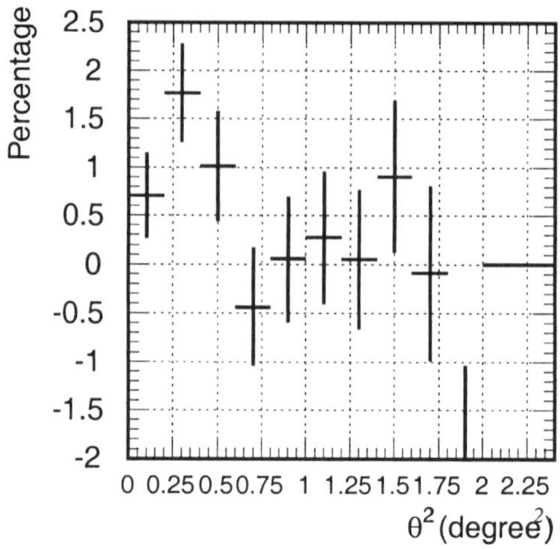

Fig. 4. Wide-range θ^2 plot for the Vela X region [5], where θ is a space angle of an event direction from (R.A., decl.) = ($8^h35^m, -45°36'$) [J2000], i.e., the peak of the emission detected by H.E.S.S. [19].

discriminant method to the stereo data for RX J0852.0-4622 observed in January and February 2005 using T2, T3 and T4 taken in the wobble mode. After the coarse selections, 1,129 minutes (ON) and 1,081 minutes (OFF) data were available. For the Fisher discriminant, we used six image parameters, lengths and widths, determined with each telescope independently. Finally the gamma-ray events were extracted by comparing the Fisher discriminant values between the SNR region and the background region. The excess count map is shown in Figure 5 [22]. The smoothing was carried out using the average of the center and neighboring eight pixels where the pixel size was $0.2° \times 0.2°$. The dotted circle of 1 degree from the supernova center is also shown. The strong gamma-ray emission from the NW rim is obviously seen, which was first reported by CANGAROO-II [20]. The emission profile shows shell-like structure like that seen in X-rays. The differential energy spectra for the whole remnant are shown in Figure 6. The agreement of the CANGAROO-III result with that of H.E.S.S. [21] is reasonable. The energy spectrum around the NW-rim was measured to be consistent with that of the whole remnant, i.e., flatter than that reported from the previous CANGAROO-II data. The difference can be partially explained by the deterioration of the hardware of the CANGAROO-II telescope.

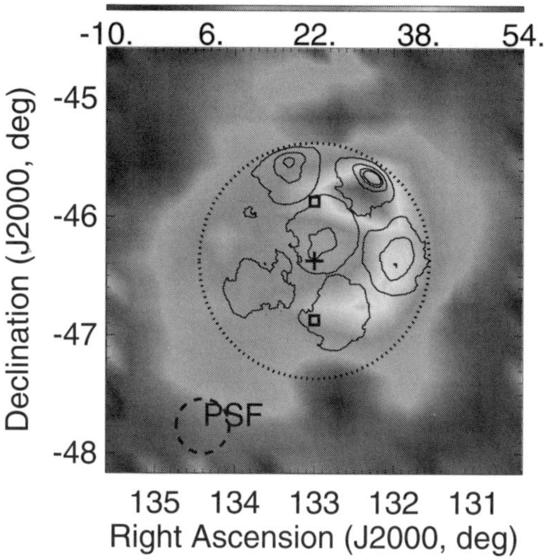

Fig. 5. Excess event map around the SNR RX J0852.0-4622 obtained from the CANGAROO-III stereo observations in 2005 [22]. The vertical scale (number of events per pixel, $0.2° \times 0.2°$) is indicated in the top bar. The cross indicates the average pointing position, i.e., the center of the remnant and the squares the "wobble" pointing positions. The dashed circle of 0.23° shows the (1σ) point spread function. Overlaid contours show X-ray intensity observed by ASCA GIS.

4. Summary

We have been carrying out stereo observations of sub-TeV gamma-rays with CANGAROO-III since March 2004. Results from stereo observations were presented: PSR 1706-44 and SN1006, from which gamma-ray signals were reported by CANGAROO-I, were not confirmed by CANGAROO-III observations. For two supernova remnants, the Vela SNR and RX J0852.0-4622, our results are consistent with the recent H.E.S.S. results.

The distribution of 'gamma-ray SNRs' is important in the quest for the origin of cosmic-rays and the high-energy content of the Universe: we will continue systematic study of SNRs in high-energy gamma-rays in the Galaxy.

References

1. Masaki Mori, *Science with the New Generation of High Energy Gamma-ray Experiments*, eds. A. De Angelis and O. Mansutti (World Scientific, Singapore, 2006), pp.21-28 and references therein.

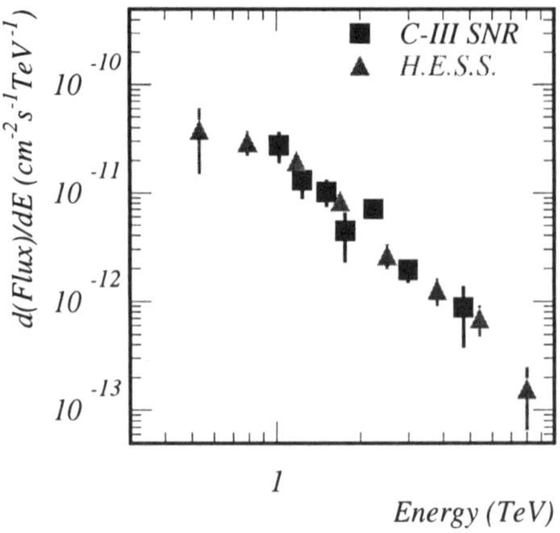

Fig. 6. Differential energy spectra; squares are data points obtained by CANGAROO-III observation for the whole remnant [22], and triangles are by H.E.S.S. [21]. The error bars are statistical.

2. T. Nakamori et al., in *Proc. 29th ICRC (Pune)* (Tata Inst. Fundamental Research, India, 2005), Vol. 4, pp. 203–206.
3. R. Enomoto et al., in *Proc. 27th ICRC (Hamburg)* (Copernicus Gesellshaft, Germany, 2001), Vol.5, pp. 2477–2480.
4. R. Enomoto et al., *Nature* **416**, 823–826 (2002).
5. R. Enomoto et al., *Astrophys. J.* **638**, 397–408 (2006).
6. F.A. Aharonian et al., *Astrophys. J.* **539**, 317–324 (2000).
7. A.M. Hillas et al., *Astrophys. J.* **503**, 744–759 (1998).
8. T. Kifune et al., *Astrophys. J.* **438**, L91–94 (1995).
9. P.M. Chadwick et al., *Astropart. Phys.* **9**, 131–136 (1998).
10. F.A. Aharonian et al., *Astron. Astrophys.* **432**, L9–L12 (2005).
11. T. Tanimori et al., in *Proc. 29th ICRC (Pune)* (Tata Inst. Fundamental Research, India, 2005), Vol. 4, pp. 215–218.
12. K. Koyama et al., *Nature*, **278**, 255–228 (1995).
13. T. Tanimori et al., *Astrophys. J.* **497**, L25–28 (1998).
14. F.A. Aharonian et al., *Astron. Astrophys.* **437**, 135–139 (2005).
15. V. Vitale et al., in *Proc. 28th ICRC (Tsukuba)* (Universal Academy Press, Tokyo, 2003), pp. 2889–2892.
16. T. Yoshikoshi et al., *Astrophys. J.* **487**, L65–68 (1997).
17. P.M. Chadwick et al., *Astrophys. J.* **537**, 414–421 (2000).
18. S.A. Dazeley, Ph.D. thesis, University of Adelaide (1999).
19. F.A. Aharonian et al., *Astron. Astrophys.* **448**, L43–47 (2006).
20. H. Katagiri et al., *Astrophys. J.* **619**, L163–L165 (2005).
21. F. Aharonian et al., *Astron. Astrophys.* **437**, L7–10 (2005).
22. R. Enomoto et al., *Astrophys. J.*, **653**, in press (2006).

The H.E.S.S. project

C. MASTERSON for the H.E.S.S. collaboration

Max-Planck-Institut für Kernphysik, P.O. Box 103980, D 69029 Heidelberg, Germany
** E-mail: conor.masterson@mpi-hd.mpg.de*
www.mpi-hd.mpg.de/hfm/HESS

The H.E.S.S. Imaging Atmospheric Cherenkov Telescope Array is currently the most sensitive instrument for Very High Energy (VHE) γ-ray observations in the energy range of about 0.1-10 TeV. While the detector is well suited for studies of extended galactic sources, an extensive program of observations of variable point-sources has also been carried out in the past three years of operation. Many galactic and extragalactic VHE γ-ray sources have been discovered, and a selection of recent H.E.S.S. results is presented in this proceeding.

Keywords: gamma rays; observations

1. The Cherenkov technique

The study of very high energy (VHE) γ-rays using ground based telescopes depends on the detection of the cascade of charged particles created by the interaction of the γ-ray within the Earth's atmosphere. The electrons and positrons produced combine to emit a short ($\sim$ 5 nsec) flash of Cherenkov light, which is detected by an array of large optical telescopes on the ground. The use of multiple detectors allows for accurate determination of the shower core position and the trajectory of the original γ-ray photon.[1]

The H.E.S.S. array is situated in the Khomas highland of Namibia, at an elevation of 1800 metres above sea level. The H.E.S.S. detector is the most sensitive of its kind in the world, capable of detecting a flux at a level of 1% of the Crab nebula (a standard source) in 25 hours. The four telescopes are placed in a square formation with a side length of 120 metres. The telescopes, of steel construction, have an effective mirror area of 107 m^2. The H.E.S.S. cameras each consist of a hexagonal array of 960 photomultiplier tubes (PMTs). Each PMT covers an area of 0.16° in diameter projected onto the sky. The total field of view (fov) on the sky is 5° in diameter. The detector has an angular resolution of 0.1° and an energy resolution of 15%. Details of the data analysis procedures for H.E.S.S. data may be found in Ref. 1 and Ref. 2.

Concurrently with the scientific observation program, phase II of the H.E.S.S. project is also under development, with the design and construction of a single very large telescope for the centre of the array. This will have a dish diameter of 35 metres and a 3° fov, with 2000 pixels. Phase II is expected to be completed in early 2008.

2. Observations of Galactic sources with H.E.S.S.

The driving strategy of the H.E.S.S. Galactic observation program has been the survey of the inner regions of the Galactic disk. Many candidate VHE γ-ray sources exist in this part of the Galaxy. These potential accelerators include the Galactic Center, supernova remnants (SNR), pulsar wind nebulae (PWN) and massive stars. Although 91 SNRs and 381 pulsars are known to exist[3,4] the central 60° in galactic longitude (l) and 6° in galactic latitude (b), only two objects (the Galactic Center[5] and the SNR RX J1713.7-3946[6]) had been detected at VHE energies in this region prior to the commencement of H.E.S.S. observations. The unprecedented sensitivity of H.E.S.S. coupled with the high density of potential VHE sources motivated a deep (230 live-hours after data-quality selection) survey[7,8] of this inner region in 2004. This survey, spanning from $-30°$ to $30°$ in longitude, and $-3°$ to $3°$ in latitude, reached an average sensitivity of 2% of the Crab nebula. Figure 1 shows a significance map of the first H.E.S.S. survey region. A total of 14 new sources have been discovered in the H.E.S.S. galactic survey.

2.1. *Supernova remnants*

Supernova remnants, left behind by gigantic explosions of massive stars, have since long been thought to be one of the prime sources of VHE cosmic rays. At least 5 sources seen in the H.E.S.S. Galactic Plane scan could be identified as SNRs. One of the best studied examples is RXJ 1713.7-3946, a shell-type supernova remnant, which was observed by H.E.S.S. in 2003. With an angular resolution more than an order of magnitude better than the spatial extension of the SNR, H.E.S.S. was the first to resolve the morphology of a TeV γ-ray source.[9] In 2004, RXJ 1713.7-3946 was re-observed for 33 hours live time. A strong signal is observed with a significance of about 39σ (Fig. 2), which enables detailed spectral and morphological studies.[10] While the H.E.S.S. results seem to favour a hadronic acceleration scenario,

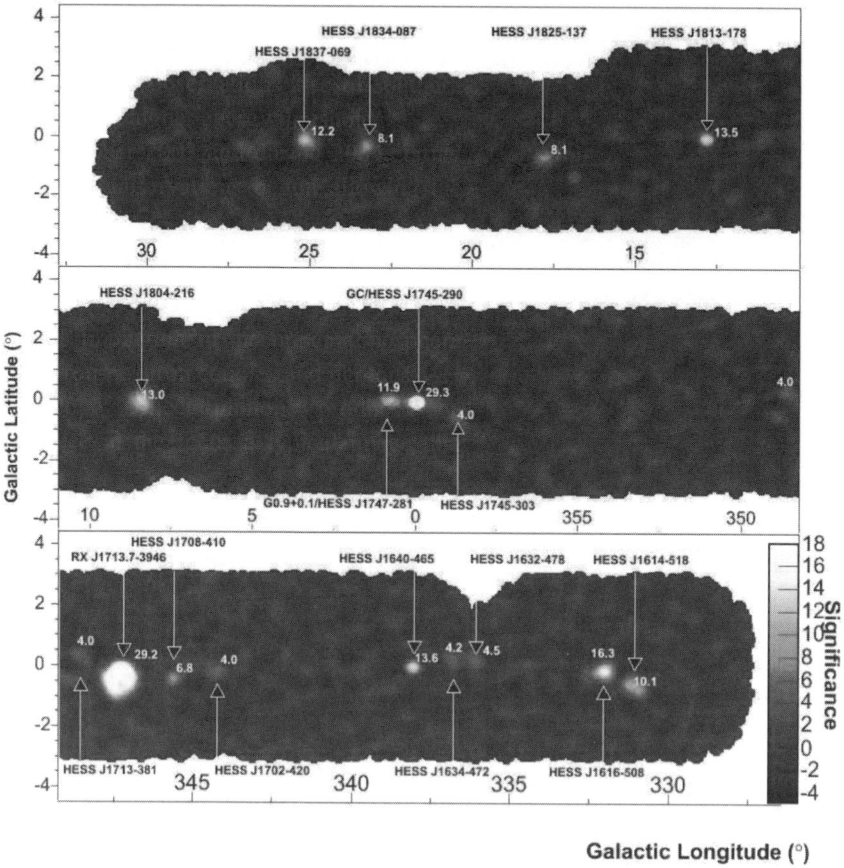

Fig. 1. Significance map of the H.E.S.S. Galactic Plane survey in 2004, including data from re-observations of source candidates from the original scan and observations of the known γ-ray sources RX J1713.7-3946 and the Galactic Centre region.

with proton-proton interactions producing the observed γ-rays, no strong conclusion can yet be drawn about the dominant acceleration mechanism involved.

2.2. *Galactic Centre region*

The inner 100 parsec of our Galaxy are known as the most violent and active region in our solar neighbourhood. Not only does it host a central supermassive black hole, but also numerous other objects like supernova remnants, massive stars and giant molecular clouds.

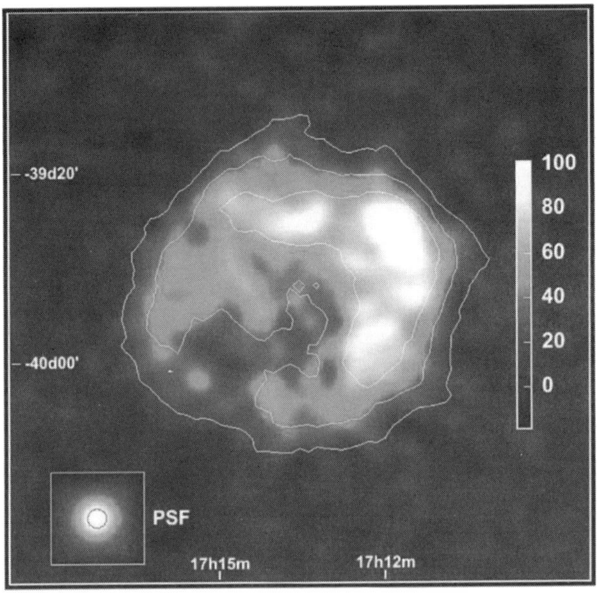

Fig. 2. Image of RXJ 1713.7-3946 in VHE γ-rays. The linear colour scale is in units of excess counts. The white contours denote the significance of the features. The levels are linearly spaced and correspond to 5, 10, and 15σ, respectively. In the lower left corner a simulated point source is shown as it would appear in this particular data set.

H.E.S.S. have observed a strong source of TeV γ-rays (HESS J1745-209) from the direction of the Galactic Centre.[11] The centre of gravity of the (almost) point-like excess is spatially coincident ($3'' \pm 12''(stat.)$) with the central black hole Sagittarius A*. However, even with the good pointing accuracy of the instrument (20"), the SNR Sgr A East cannot be ruled out as the source of the observed emission. Figure 3a shows the region surrounding the Galactic centre, including the source coincident with Sgr A* and the nearby source coincident with the plerion G0.9-0.1. In figure 3b these sources are subtracted, showing a region of diffuse emission coincident with dense molecular clouds close to the Galactic centre.[12]

3. Observations of Active Galactic Nuclei

Several nearby active galactic nuclei (AGN) of the "blazar"-type are known to emit VHE (E >100 GeV) γ-rays. Understanding and modelling their emission through spectral and variability studies can lead to improved understanding of the emission processes involved, but it is vital to take into ac-

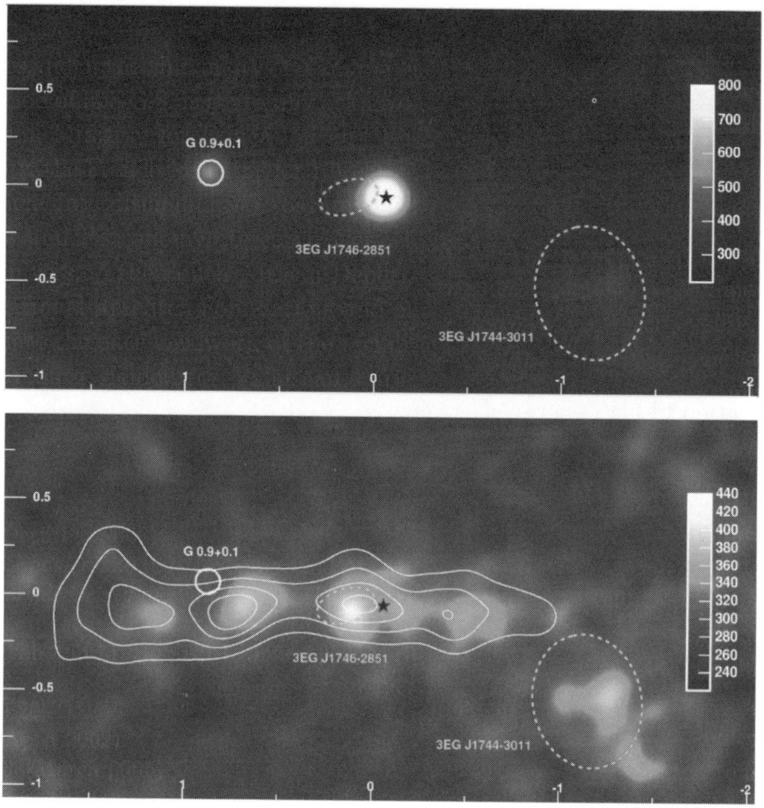

Fig. 3. VHE γ-ray images of the GC region. (a) γ-ray count map, showing the strong source G0.9+0.1 and the emission coincident with the gravitational centre of our galaxy. The position of Sagittarius A* is marked with a black star. (b) Same map after subtraction of the two bright sources. White contour lines indicate the density of dense molecular clouds, traced by CS emission. The green ellipses show 95% confidence regions for the position of two unidentified EGRET sources.

count their emission across all available wavelengths. Currently, constraining any emission model is difficult as only a limited number of high-energy measurements currently exist. Extragalactic γ-ray emission, especially from AGN, can be used as a probe of the extragalactic background light, as γ-rays interact with optical and near-IR photons, the distribution of which is difficult to measure at these distances citeebl06.

3.1. *PKS 2155*

PKS 2155−304 (redshift $z = 0.116$) has been detected with high significance ($\sim 56\sigma$) using H.E.S.S., confirming a previous VHE detection.[13] A strong signal is found in each of the data sets corresponding to the dark periods of July and October, 2002, and June-November, 2003. A first publication[14] using data through September 2003 reported the initial detection ($\sim 45\sigma$) of PKS 2155−304. Variability in the observed flux of VHE γ-rays from PKS 2155−304 is found on time scales of months, days, and hours. The spectral energy distribution of the emission from this source is shown in figure 4, with the characteristic double-peaked structure of the emission, possibly caused by synchrotron and inverse Compton interactions of a single population of high energy electrons.

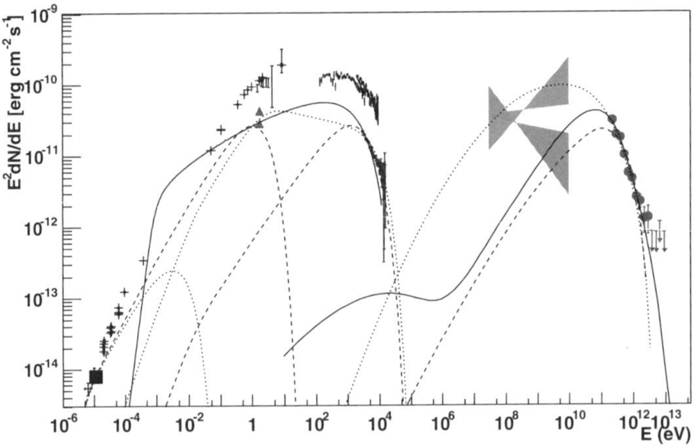

Fig. 4. The SED of PKS 2155−304. Only simultaneous observations are labeled. Non-contemporaneous archival data are shown in grey. The solid line represents a hadronic model, and the dotted and dashed lines represent variations of a leptonic model.

3.2. *Markarian 421*

Strong X-ray flares detected by the RXTE all sky monitor (ASM) triggered observations of the well-known VHE emitter Mkn 421 ($z = 0.03$) with the H.E.S.S. telescopes from January through May 2004. Observations could only be made at zenith angles larger than 60° due to the northern declination of the object ($\sim 38°$), which results in an relatively large energy threshold for these data (~ 1.5 TeV). In total, 14.7 hrs live time (mainly in

April 2004) of observations were made. Overall, an excess of ~7600 events, corresponding to a significance of 114σ, was measured from the direction of Mkn 421.[15] The measured time-averaged spectrum is shown in Fig. 5 and is clearly curved. The average integral flux observed (>2 TeV) is ~3 times higher than the flux measured from the Crab Nebula and as can be seen in Figure 5 exhibits strong variability on a nightly basis. The maximum amplitude of the variations reaches 4.3 ± 0.5. The inlay of Fig. 5 shows the night with the strongest indication for intra-night variability, where the flux drops by a factor of 2 within one hour. Additionally, significant variations in the spectral shape, best described as a hardening with increased flux, are found.

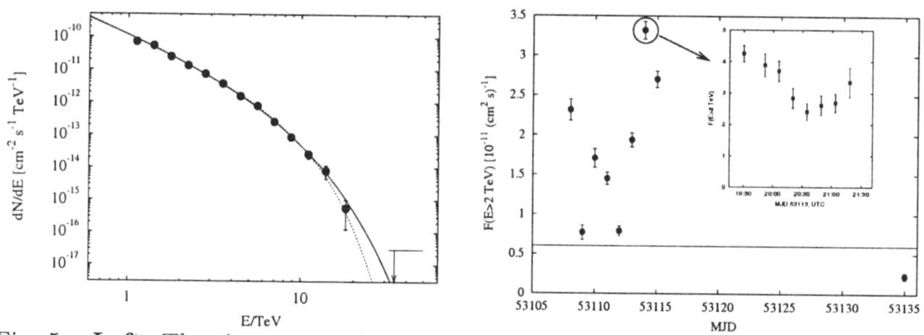

Fig. 5. **Left:** The time-averaged energy spectrum of Mkn 421. The solid line is the best fit of a power law ($\Gamma = 2.1 \pm 0.1_{\rm stat} \pm 0.4_{\rm syst}$) with an exponential cutoff ($3.1(+0.5 -0.4)_{\rm stat} \pm 0.9_{\rm syst}$ TeV), and the dashed line is the best fit of a super-exponential cutoff. **Right:** The nightly integral flux (>2 TeV) observed from Mkn 421. The solid line is the Crab Nebula flux. The inlay shows the variability detected during MJD 53113 (14 min bins).

4. Conclusion

The H.E.S.S. experiment has been fully operational with four telescopes since January of 2004 and has achieved its designed sensitivity and operational characteristics. Useful scientific results have been produced throughout the construction phase with two three and four telescopes and exciting discoveries have been made. An extensive campaign of observations of galactic sources is underway and more exciting results are expected. Sources discussed here will be re-observed with the improved sensitivity and resolution of the full array and such deeper observations are expected to provide further interesting insights from these sources.

5. Acknowledgments

The support of the Namibian authorities and of the University of Namibia in facilitating the construction and operation of H.E.S.S. is gratefully acknowledged, as is the support by the German Ministry for Education and Research (BMBF), the Max Planck Society, the French Ministry for Research, the CNRS-IN2P3 and the Astroparticle Interdisciplinary Programme of the CNRS, the U.K. Particle Physics and Astronomy Research Council (PPARC), the IPNP of the Charles University, the South African Department of Science and Technology and National Research Foundation, and by the University of Namibia. We appreciate the excellent work of the technical support staff in Berlin, Durham, Hamburg, Heidelberg, Palaiseau, Paris, Saclay, and in Namibia in the construction and operation of the equipment.

References

1. F. Aharonian *et al.*, *A&A* **457**, 899(Octtober 2006).
2. F. A. Aharonian *et al.*, *Astroparticle Physics* **22**, 109(November 2004).
3. D. A. Green, *Bulletin of the Astronomical Society of India* **32**, 335(December 2004).
4. R. N. Manchester, G. B. Hobbs, A. Teoh and M. Hobbs, *AJ* **129**, 1993(April 2005).
5. K. Kosack *et al.*, *ApJ* **608**, L97(June 2004).
6. R. Enomoto *et al.*, *Nature* **416**, 823(April 2002).
7. F. Aharonian *et al.*, *Science* **307**, 1938(March 2005).
8. F. Aharonian *et al.*, *ApJ* **636**, 777(January 2006).
9. F. A. Aharonian *et al.*, *Nature* **432**, 75(November 2004).
10. F. Aharonian *et al.*, *A&A* **449**, 223(April 2006).
11. F. Aharonian *et al.*, *A&A* **425**, L13(Octtober 2004).
12. F. Aharonian *et al.*, *Nature* **439**, 695(February 2006).
13. P. M. Chadwick *et al.*, *ApJ* **513**, 161(March 1999).
14. F. Aharonian *et al.*, *A&A* **430**, 865(February 2005).
15. F. Aharonian *et al.*, *A&A* **437**, 95(July 2005).

THE MAGIC EXPERIMENT

NICOLA TURINI
For the MAGIC Collaboration

University of Siena and INFN Pisa, via Roma 56
53100 Siena, Italy

The Magic Cherenkov Telescope is located at the Observatory of the Roque de los Muchachos (ORM) and since 2004 is operative, in may 2005 was started the first yearly campaign of data taking named Cycle I. During this period the telescope reported signals from galactic and extragalactic AGN sources. In this report we will describe the technique for the detection of Very High Energy (VHE) gammas by this class of experiments and the first results from Cycle I observation.

1. The MAGIC telescope

1.1. *Introduction*

The MAGIC telescope is a single mirror instrument designed to detect the Cherenkov light produced by the electromagnetic cascades initiated by incoming high energy gammas on the upper atmosphere [1].

The high energy gammas (50Gev-10Tev) passing through high atmosphere start an electromagnetic shower from an altitude of 15 Km. The main idea is to use the atmosphere as a giant electromagnetic calorimeter to detect the energy and the incoming direction of cosmic gammas. The shower properties are reconstructed through the detection of the Cherenkov light emitted by superluminal electrons generated in the cascade.

The light pool produced by the showers covers an area of typically $10^5/10^6 m^2$, such large detection area is independent from the mirror size making the Cherenkov telescopes highly sensitive. The dish area indeed is important for the detection of the lower energy gammas.

The MAGIC telescope is actually the largest telescope of this category of instruments and therefore its energy threshold is the lowest, 50GeV, filling the gap of the covered spectrum between satellites experiments and other VHE gamma experiments.

The Telescope is located at an altitude of 2200 m in the area of the Observatorio del Roque de los Muchachos, in the Canary islands, covering mainly northern emisphere sources, but allowing observations also of southern emisphere candidates, as galactic center, with an higher energy threshold.

The experiment was inaugurated in October 2003 and during 2004 started commissioning runs. In may the Cycle I of observations begun allowing the detection of known and new sources.

1.2. *Detection principles*

The gamma primary high energy photons impinging in the upper atmosphere initiate a shower made exclusively by electron-positron couples and photons.

The large parabolic mirror telescope collect photons produced by superluminal charged particles and produce in the focus a typical image that develops in short time.

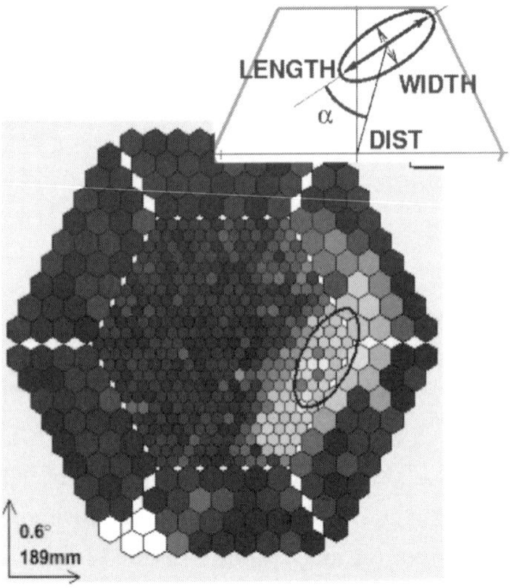

Figure 1. Typical hadron shower image on the MAGIC camera, and image parameters used to select gamma-like images.

The main source of background is generated by night sky light produced by stars and light scattered by the atmosphere. The other main source of background is given by showers initiated by different particles than gammas. In particular high energy protons have a rate three order of magnitude larger than the brightest gamma source and have a isotropic direction distribution.

The main background for the lower energy gammas is produced by large impact parameters muons that mimic gamma-like images reproducing also temporal behavior. The missing of a clear signature for low energy (<100GeV) gamma images rises the analysis energy threshold, while the trigger threshold has been measured to be well below 50GeV.

1.3. *The telescope technology*

The MAGIC telescope has been developed with the target of lowering the energy threshold of the instrument. For this reason new technologies had been developed during the R&D period that has been proved crucial. In particular three main issues has been exploited to reach low energy threshold: high quantum efficiency photosensors[2], large lightweight space frame and mirrors[3], low integration time in both trigger[4] and data acquisition[5].

The main dish has a mean diameter of 17m for an equivalent reflecting surface of 234m^2 made by 936 0.5mX0.5m square mirrors. The mirrors are made by lightweight composite aluminum structure. The reflecting surface has been obtained milling a spherical shape on the front aluminum panel.

The other goal of the experiment is to exploit observations on Gamma Ray Burst and the entire structure has to be repositioned in a short time from the alerts coming from satellites. For this reason the structure has been made with carbon fiber tubes allowing a maximum weight of 17t. The supports are made of steel tubes and the final gross weight of the telescope is 65t.

Since the structure is not enough stiff for different zenith angles the mirror panels are focused dynamically with an Active Mirror Control system. Each panel has a laser pointer that allows to measure the off pointing and a dynamical system, based on Look Up Tables, continuously adjust each panel for different zenith angles.

The photosensors are 1 inch photomultipliers for the inner pixels and 1.5 inches tubes for the external low resolution pixels. The tubes have a semispherical entrance window to enlarge the probability of double hit of the incoming photons and a wavelength shifter coating has been added on the surface that increased the mean quantum efficiency by 30% from the original one. The same coating allowed sensitivity on the UV band as shown in fig2.

To preserve the signal integrity the transmission has been made with an analog optical system. A VCSEL laser diode transmits the signals over 175m from the telescope camera to the counting house, there fast receiver boards converts back to electrical the information and prepare the signals for the trigger and data acquisition.

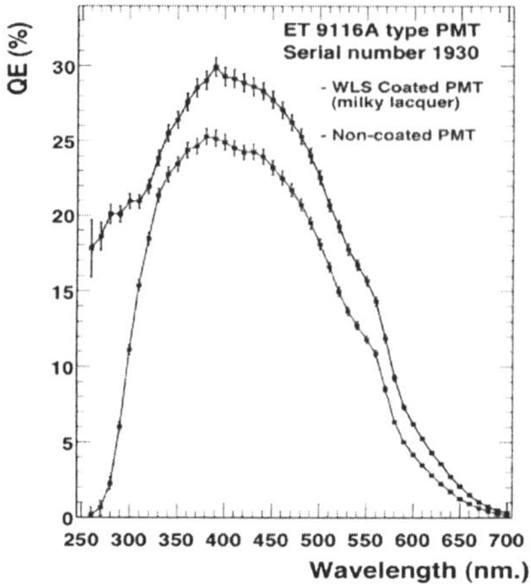

Figure 2. Quantum Efficiency measurements on 1 inch Electron Tube 9116A photomultiplier. lower non coated PM, upper coated PM.

To reduce at maximum the NSB noise, the trigger performs coincidences in windows of 6ns on selectable close compact images. The DAQ is made by 300MHz samplers that acquire a time window of 30ns to fully reconstruct the original signal. Since the sampling frequency is a little marginal compared with the expected width of the shower signals, a filter has been added to stretch up to 6 ns the peaks to allow a good measurement of the absolute arrival time of the light on the tube. A double gain system has been added to increase the dynamical range of the 8 bit FADC of the samplers.

2. Cycle 1 observations

Introduction

From may 2005 the telescope started an observation campaign, named Cycle1, but many hours of observations on galactic and extragalactic sources where already available from commissioning phase. The data taken since the last months of 2004, where the telescope performed almost on it's design expectations are included in the analysis presented here.

During Cycle I where taken 750 hours of data from extragalactic sources and 500 hours from galactic sources, apart from Crab used as standard candle to calibrate the instrument.

In this phase we have monitored the performances of the telescope and measured the energy and angular resolution. The measurements show an energy resolution of 20% for higher energies increasing up to 35% for threshold energies, while the spatial resolution show a PSF of ~0.1° for higher energies increasing to ~0.13° for threshold energies. The Crab plerion has been observed during winter 2004 and 2005 revealing a gamma signal at the telescope of ~20σ/h

2.1. *Extragalactic sources*

The search for very high energy γ-ray emission from Active Galactic Nuclei (AGNs) is one of the major goals for ground-based γ-ray astronomy,

The photon-photon interaction produce a cut-off for higher energies and a correlation between redshift and maximum energy of detected sources, known as Fazio-Stecker relation, is expected.

The Cycle I strategy on this category of sources was to observe a sample of X-ray bright (F_{1keV} >2 μJy) northern HBLs at moderate redshifts (z<0.3). The sample of candidates was chosen based on predictions from models involving an SSC and hadronic origin of the γ-rays

In parallel with the observations of the AGNs with MAGIC, the sources were observed with the KVA 35 cm telescope, also located on La Palma. In addition to these joint campaigns, several AGNs are regularly observed as part of the Tuorla Observatory Blazar monitoring program with the Tuorla 1 m and the KVA 35 cm telescopes.

Table 1. Summary of observation of Cycle I AGN survey. ZA is the zenith angle in degrees. E_{Thr} is the analysis threshold in GeV. T(hr) is the observation time in hours.

Source	period	ZA	E_{Thr}	T(hr)
Mkn 421	Nov04-Apr05	9-55	150	25.6
Mkr 501	Jun-Jul05	10-31	150	29.7
1ES 2344	Sep-Dec05	23-34	160	27.6
Mkn 180	Mar06	39-44	200	11.1
1ES 1959	Sep-Oct04	36-46	180	6.0
1ES 1218	Jan05	2-13	140	8.2
PG 1553	Apr-May05 Jan-Apr06	12-30 2030	140	7.0 11.8

During the entire observation period Mkn 421 [6] was found to be in a low to moderate high flux state ranging from 0.5 to 2 Crab units above 200 GeV. Significant variations of up to a factor of four overall and up to a factor two in between successive nights has been seen and a clear correlation between X-ray (taken from the All-Sky-Monitor on-board the RXTE satellite) and VHE γ-ray data is verified.

Mkn 501 was observed for 24 nights (Table 1) by the MAGIC telescope. The source was found to be in a rather low flux state during most of the observations. The flux level above 200 GeV was 30%-50% of the Crab Nebula flux with a strong indication of an IC peak. On five nights, the source was found in a flaring state with the flux reaching up to 4 Crab units Moreover, a rapid flare with a doubling time as short as 5 minutes or less was detected on the night of 10 July 2005. The rapid increase in the flux level was accompanied by a hardening of the differential spectrum. This is the first time that spectral hardening was detected on time scales of some 10 minutes. A detailed publication on the analysis and results on the observation of Mkn 501 is in preparation.

1ES 2344+514, reported a VHE γ-rays emission by Wipple. in 1996. The MAGIC observation of 1ES 2344+514 (Table 1) yielded a clear excess with the significance of 11.5σ. The differential energy spectrum can be fitted by a simple power law with a photon index of 2.96 ± 0.12.

The AGN Mkn 180[7] (1ES 1133+704) is a well-known HBL at a redshift of $z = 0.045$. The observation of Mkn 180 was triggered by a brightening of the source in the optical on March 23, 2006, detected by the KVA telescope. The alert was given as the core flux increased by 50% from its quiescent level value. Mkn 180 was observed by the MAGIC telescope in 2006 during 8 nights (Table 1). The signal of 165 excess events was found with a significance of 5.5 σ. No evidence for flux variability was found.. The observed integral flux above 200 GeV is 11% of the Crab Nebula flux.

1ES 1959+650[8] was observed by MAGIC during the commissioning phase in 2004 (Table 1). The analysis of the data gave a detection of VHE γ-rays on a significance of 8.2σ

MAGIC observed 1ES 1218+304[9] in seven nights in January 2005 (Table 1). The observed excess of 560 events has a statistical significance of 6.4 standard deviations above 140 GeV. 1ES 1218+304 is the first source discovered by MAGIC. The gamma-ray lightcurve did not show signs of significant variability. The energy spectrum was fitted with a pure power law with a photon index of 3.0 ± 0.4

PG 1553+113[10] was observed with the MAGIC telescope in 2005 at about the same time as the H.E.S.S. observations took place. Motivated by a hint of a signal in the preliminary analysis of the MAGIC data, additional observations were performed in 2006 (Table 1). Combining the data from 2005 and 2006, a very clear signal was detected with a total significance of 8.8 σ.

2.2. *Galactic sources*

About 1/3 of the observation time (not counting that devoted to Crab nebula technical observations) was devoted to galactic objects. The observations covered both candidates and well established VHE γ-ray emitters, and included the following types of objects: supernova remnants (SNRs), pulsars, pulsar wind nebulae (PWN), microquasars (μQSRs), the Galactic Center (GC), one unidentified TeV source and one cataclysmic variable. In this section we highlight the results obtained so far from such observations.

We have confirmed the VHE γ-ray emission from the SNRs HESS J1813-178[11] and HESS J1834-087 (W41)[12].

We have also measured the VHE γ-ray flux from the GC[13]. The possibility to indirectly detect dark matter through its annihilation into VHE γ-rays has risen the interest to observe this region during the last years. Our observations have confirmed a point-like γ-ray excess whose location is spatially consistent with Sgr A* as well as Sgr A East. The energy spectrum of the detected emission is well described by an unbroken power law of photon index α= -2.2 and intensity about 10% of that of the Crab nebula flux at 1 TeV.

LS I +61 303[14] was observed in the VHE regime with MAGIC during 54 hours (after standard quality selection, discarding bad weather data) between October 2005 and March 2006

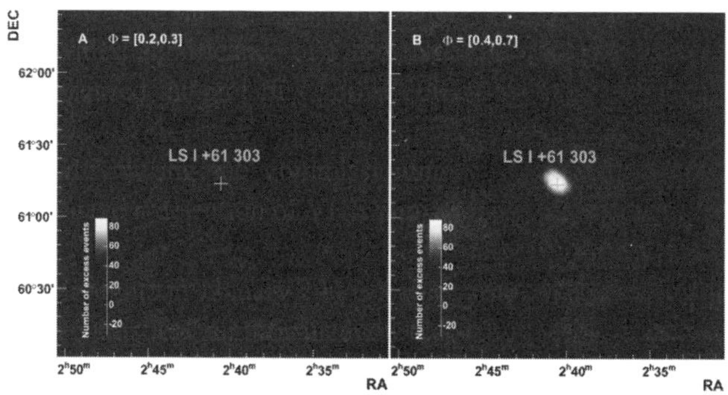

Figure 3. Smoothed maps of γ-ray excess events above 400 GeV around LS I +61 303. (A) Observations over 15.5 hours corresponding to data around periastron (i.e., between orbital phases 0.2 and 0.3). (B) Observations over 10.7 hours at orbital phase between 0.4 and 0.7.

The microquasar shows an evident pulsed VHE emission in phase with orbital parameters. The maximum of the pulse is present between phase 0.4-0.7, while at periastron is absent. This is the first periodic signal seen from VHE emitters.

3. Conclusions

The magic telescope has demonstrated to work within the design specification, the analysis on the low energy gammas on the other hand has been proved difficult due to the background. Actually we are upgrading the experiment adding a mechanical clone of the MAGIC, named MAGICII. The stereo images are proven to reduce the muon background and improve the energy and direction resolution. The two telescopes design is foreseen to double the sensitivity.

The new telescope will include many new technologies as lighter 1 square meter mirrors, new photosensors and new faster samplers. The big improvement is expected by high quantum efficiency devices such as HPD's and SiPM. A new 2GHz sampler, based on a custom chip named Domino, is foreseen to reduce the NSB contribution on the signal. The heat dissipation required by standard FADC is causing troubles and the new design will reduce it by 2 orders of magnitude.

The MAGICII telescope is expected to start it's commissioning phase late 2007 and join MAGIC in the data taking in 2008.

Acknowledgments

We would like to thank the IAC for the excellent working conditions at the ORM in La Palma. The support of the German BMBF and MPG, the Italian INFN, the Spanish CICYT, ETH research grant TH 34/04 3, and the Polish MNiI grant 1P03D01028 is gratefully acknowledged

References

1. A. Moralejo, Memorie delle Societa Astronomica Italiana, v. 75, p. 232 (2004).
2. D. Paneque et al., Nucl.Instrum.Meth. A 518 (2004) 619.
3. C. Bigongiari et al., Nucl.Instrum.Meth. A 518 (2004) 193.
4. M. Meucci, R. Paoletti et al., NIM A (2004), Issue 1-2, 554.
5. H.Bartko et al., astro-ph/0506459
6. J. Albert et al,. submitted to ApJ 2006
7. J.Albert et al,.submitted to ApJ Letters in June 2006
8. J.Albert et al., ApJ 639 (2006) 761-765
9. J. Albert et al., ApJ Letters 642, L119 (2006)
10. J.Albert et al,.submitted to ApJ Letters in May 2006
11. J. Albert et al., ApJ Letters 637, L41 (2006)
12. J. Albert et al., ApJ Letters 643, L53 (2006)
13. J. Albert et al., ApJ Letters 638, L101 (2006)
14. J. Albert et al., Science 312, 1771 (2006)

VERITAS: Status and Performance

J. Holder* for the VERITAS collaboration

Department of Physics and Astronomy, University of Delaware,
Newark, DE 19716, USA
**E-mail: jholder@physics.udel.edu*

VERITAS is an atmospheric Cherenkov telescope array sited in Tucson, Arizona. The array is nearing completion and consists of four, 12 m diameter telescopes. The first telescope in the array has been operating since February 2005, while observations with the full array are expected to begin in January, 2007. We report here in some detail on the performance of the first VERITAS telescope, and briefly discuss the first stereo observations.

Keywords: Gamma Ray Astronomy, Cherenkov Telescopes

1. Introduction

The Whipple 10 m telescope provided the first detection of an astrophysical γ-ray source, the Crab Nebula, using the technique of forming images of the atmospheric Cherenkov emission generated by air showers.[1] Arrays of imaging atmospheric Cherenkov telescopes provide a further increase in sensitivity and in angular and energy resolution, as demonstrated by the HEGRA experiment.[2] The stereoscopic imaging atmospheric Cherenkov technique is now being exploited by four major atmospheric Cherenkov observatories; HESS,[3] MAGIC,[4] CANGAROO III[5] and VERITAS.[6] The VERITAS array comprises four, 12 m telescopes and will be used to observe astrophysical sources from the northern hemisphere over the energy range from 100 GeV to 50 TeV, with a sensitivity of 7 mCrab (a gamma-ray source of 0.7% of the Crab Nebula flux will be detected with 5σ significance over a 50 hour exposure). The first telescope came online in February 2005,[7] followed by the first stereo observations with two telescopes in January 2006. We present here a detailed description of the performance of the first telescope and some initial results from the stereo system.

2. The First VERITAS Telescope

Figure 1 shows the first VERITAS telescope installed at the Whipple observatory basecamp in Tucson, Arizona, at an altitude of 1275 m above sea level. The 12 m diameter collector is used to produce an image of the Cherenkov emission generated by gamma-ray and cosmic-ray air showers on a 499 photomultiplier tube (PMT) camera at the focus. The buildings in the foreground house the trigger and data acquisition electronics and the power supply systems.

Fig. 1. The first VERITAS telescope installed at the Whipple Observatory basecamp

2.1. *Optical and Mechanical Performance*

The telescope optics consist of a tessellated reflector of Davies-Cotton design[8] with 12 m diameter and a focal length of 12 m. The reflector is comprised of 350 individual facets, each with an area of 0.32 m^2, giving a total mirror area of $\sim$ 110 m^2. The facets are made from glass, which is slumped and polished, then cleaned, aluminized and anodized at an on-site optical coating laboratory. The reflectivity of the coating is typically $> 90\%$ at 320 nm. Reflector facets are aligned manually using a laser alignment system located at a distance of twice the focal length (24 m) from the centre of the reflector. The point spread function response is shown in Figure 2 and is measured to be $\sim 0.06°$ (full width at half-maximum) at the position of Polaris (elevation 31°).

The mirrors are distributed over a tubular steel optical support structure which is mounted on a altitude-over-azimuth positioner. The maximum slew speed of the positioner is measured to be $\sim 1^\circ\ s^{-1}$. Figure 2 shows the encoder measurements for a short tracking run, illustrating that the tracking is stable with a relative pointing accuracy of typically $< \pm 0.01^\circ$.

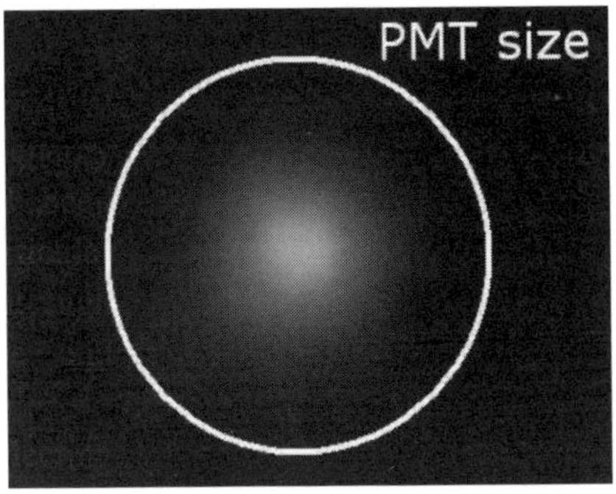

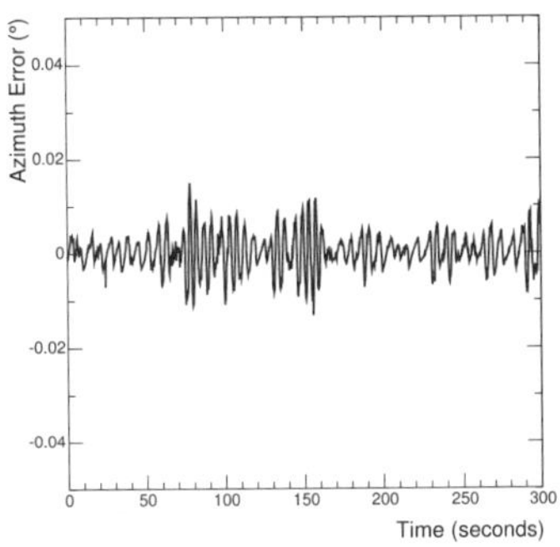

Fig. 2. **Left:** An image of Polaris in the focal plane recorded with a CCD camera. The circle indicates the size of a PMT (0.15°). **Right:** The Azimuth encoder residuals (difference between measured and requested position) for a short tracking run.

2.2. *The Camera and Electronics*

The imaging camera, shown in Figure 3, consists of 499, 2.86 cm diameter PMT pixels (Photonis XP2970/02). The pixel spacing corresponds to an angular separation of 0.15°, giving a total field-of-view of $\sim 3.5^\circ$ diameter. Reflecting light cones have now been installed on the camera face and increase the overall photon collection efficiency by $\sim 30\%$. The PMTs are operated at a gain of $\sim 2 \times 10^5$, giving typical anode currents of 3 μA (for dark fields) to 6 μA (for bright fields). Signals from the PMTs are amplified by a high-bandwidth pre-amplifier integrated into the PMT base mounts and then sent via coaxial cable to the telescope trigger and data acquisition electronics housed in the control room. The trigger electronics have two levels; each channel is equipped with a constant fraction discriminator (CFD) which produces an output logic pulse of programmable width (typically 10 ns) when the discriminator threshold is crossed.[9] These signals are then passed to a topological trigger system, similar to that used successfully on the Whipple 10m telescope,[10] which is used to detect pre-programmed patterns of triggered pixels in the camera (for example, any three adjacent pixels). The topological trigger system greatly reduces the rate of triggers due to random fluctuations of the night sky background light and enabled us to operate the first telescope alone with a CFD threshold of $\sim$ 7 photoelectrons and a cosmic ray trigger rate of $\sim$ 200 Hz at high elevation.

In the event of a trigger, each PMT signal is digitized by a custom built, 8-bit Flash-ADC system, sampling at 500MHz with a memory depth of 32 μs.[11] The FADC readout window and size are programmable; Figure 3 shows a typical FADC trace with a readout window of 48 ns (24 FADC samples). The use of FADCs, as opposed to simple charge integrating ADCs, allows us to investigate various digital signal processing techniques to improve the signal/noise ratio in each channel, as well as providing information about the time structure of the Cherenkov image in the camera, which may be useful as an additional tool for shower parameter reconstruction and gamma-hadron discrimination.[12]

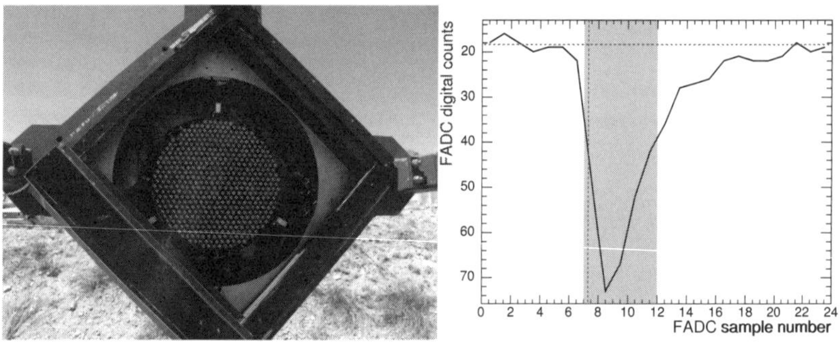

Fig. 3. **Left:** The 499 pixel imaging camera. The focus box is 1.8 m square. Light cones have now been installed. **Right:** A FADC trace resulting from Cherenkov light in a single PMT. The shaded area indicates a 10 ns (5 FADC samples) integration window.

2.3. *Data Analysis and Results*

Prior to analysis, the signal channels must be calibrated. Relative (pixel-to-pixel) gain calibration is achieved by uniformly illuminating the camera face with a $\sim$ 4 ns pulse of laser light via an opal diffuser. Electronic pedestal offsets are measured by force-triggering the data acquisition at a rate of 3 Hz during observations. Absolute calibration, which is necessary to understand the overall collection efficiency of the telescope, is achieved by measurements of the camera response to single photoelectron level signals[7] and to the Cherenkov light produced by local muons.

Following calibration, images are cleaned using a two-pass method. In the first pass, a wide (20 ns) integration window is applied to each FADC trace in order to calculate the integrated charge and the pulse arrival time.

PMTs with a clear signal are selected and the resulting image parameterised with a second moment analysis, the results of which can be described by an ellipse as shown in Figure 4. Using the FADC information, the time gradient across the image can also be measured, and this information is used in a second pass over the image in order to position a shorter (10 ns) charge integration window according to the expected position of the pulse.

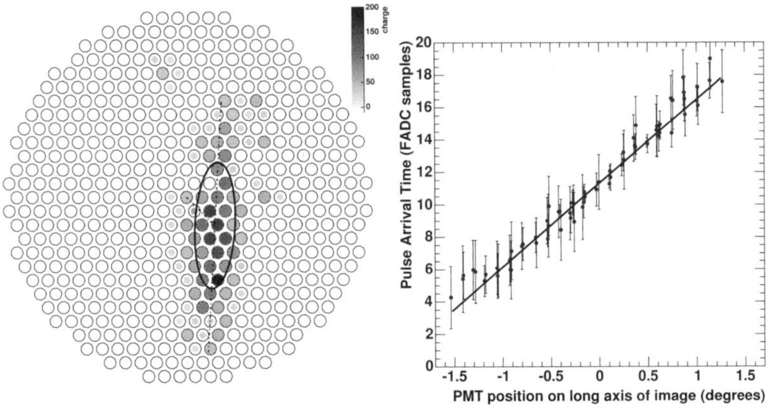

Fig. 4. **Left:** The charge distribution across the camera for a cosmic ray event. **Right:** The pulse arrival time distribution along the long axis of the ellipse for the cosmic ray image on the left.

The standard candle of ground-based gamma-ray astronomy is the Crab Nebula, and observations of this object were made in order to test the performance of the first VERITAS telescope. Figure 5 shows the map of the reconstructed source position for a 3.9 hour exposure to the Crab showing the source at the centre of the field-of-view. The single telescope sensitivity to the Crab was $\sim 10\sigma/\sqrt{\text{hour}}$. Also shown in Figure 5 is the reconstructed energy spectrum of the gamma-ray flux from the Crab Nebula, as compared to earlier results. A power-law fit to the data points gives a spectral index of 2.6 ± 0.3 and a differential flux at 1 TeV of $(3.26 \pm 0.9) \cdot 10^{-7}$ m^{-2}s^{-1}TeV^{-1} (statistical errors only). The energy reconstruction relies heavily on input from Monte Carlo simulations and so the Crab spectrum reconstruction (along with image parameter comparisons[13]) provides a confirmation that the telecope efficiency and operating parameters are well modelled.

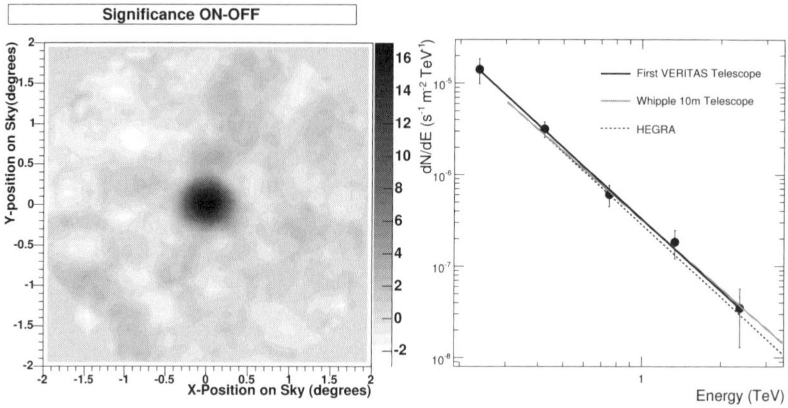

Fig. 5. **Left:** The map of reconstructed source position for observations of the Crab Nebula. The source was located at the centre of the camera. **Right:** The energy spectrum of gamma-ray emission from the Crab Nebula compared with earlier results

3. The Stereo System

The second telescope was installed on the same site and began initial operations in January 2006. For the first few months the two telescopes operated independently and coincident events were identified offline using the GPS timestamps. In March 2006 a hardware array trigger system was installed which corrects the single telescope triggers for changing time delays due to the source movement across the sky. The time-corrected signals are passed to a coincidence unit, and only events which trigger both telescopes within a coincidence time window of 100 ns are read out.

The stereo trigger serves two purposes: it allows us to run the telescopes at a lower discriminator threshold without being overwhelmed by night sky background triggers, and it removes events due to muons with impact parameters close to the individual telescopes, which provide an irreducible background for gamma-ray observations with a single telescope. Figure 6 shows both of these effects; the left-hand plot illustrates that with the hardware array trigger requirement we are able to operate the system with single telescope discriminator settings equivalent to $\sim$ 4 photoelectrons. On the right we show the ratio of image *length* to *size* (integrated charge over the whole image). Events due to local muons are distinguished by their relatively constant brightness per unit length, hence they display a distinctive peak in the single telescope histogram of *length*/*size*. When the two-telescope requirement is applied, either in hardware or in software by matching event timestamps, this peak disappears, indicating that the muon events have been removed. Note also that the hardware requirement produces a distribution with many more small size events, as the triggering threshold is significantly lower in the hardware stereo as compared to the software stereo case.

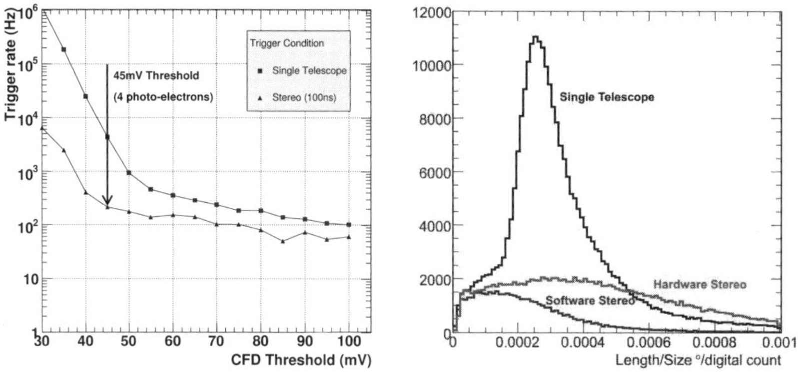

Fig. 6. **Left:** The trigger rate as a function of discriminator threshold for a single telescope and with the two telescope hardware trigger requirement. **Right:** *length/size* distributions. the single telescope histogram includes all events which trigger the telescope; the distinctive peak is due to local muons. The "software stereo" curve shows those events which remain after requiring that the events in both telescopes have a matching GPS timestamp within a 10 μs window. The "hardware stereo" curve shows the same with the two-telescope hardware trigger operating.

The two telescope system was used to observe a selection of known gamma-ray sources from March to July 2006. The TeV blazar Markarian 421 was particularly active during this period and provided a high statistics sample of TeV gamma-rays with which to test the stereo system. Figure 7 shows the distribution of θ^2 - the squared angular distance from the source location - for Mrk 421 observations. A strong source is evident.

4. Conclusion

Observations made with the first VERITAS telescope, operating since early 2005, have enabled us to verify that all technical specifications have been. met. A number of previously known gamma-ray sources, including the Crab Nebula and Mrk 421 have been detected, and the single telescope simulations have been verified by comparison with real data. Initial stereo observations have shown that the hardware array trigger system is working as expected, and that the local muon background has been removed. The construction of VERITAS is nearing completion. Figure 7 shows all four telescope structures assembled on site. Operations with the full array are scheduled to commence in January 2007, with a science program involving observations of supernova remnants, active galactic nuclei, and a survey of the galactic plane.

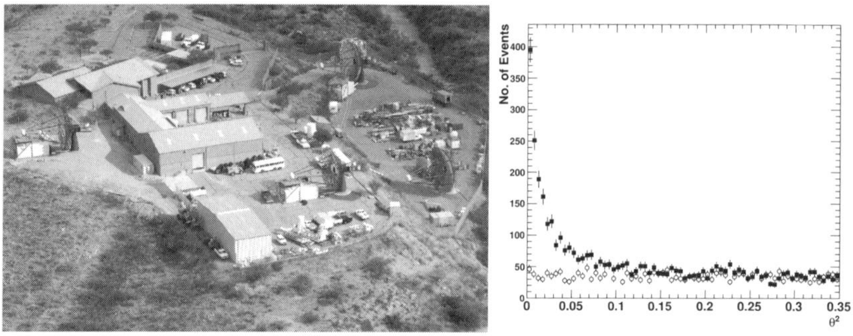

Fig. 7. **Left:** The four-telescope VERITAS array. **Right:** The θ^2 distribution for stereo observations of Markarian 421. Filled squares are observations on source, open circles are observations of an off-source control region.

5. Acknowledgments

Supported by grants from the U.S. Department of Energy, the U.S. National Science Foundation, the Smithsonian Institution, by NSERC in Canada, by Science Foundation Ireland and by PPARC in the UK.

References

1. Weekes, T. C., et al. ApJ, 342 (1989) 379
2. Pühlhofer, G. et al. Astropart. Phys., 20 (2003) 267
3. Hinton, J.A. et al. New Astron. Rev., 48 (2004) 331
4. Lorenz, E. et al. New Astron. Rev., 48 (2004) 339
5. Kubo, H. et al. New Astron. Rev., 48 (2004) 323
6. Weekes, T.C. et al. Astropart. Phys., 17 (2002) 221
7. Holder, J. et al. Astropart. Phys., 25 (2006) 391
8. Davies, J.M. & Cotton, E.S. Journal of Solar Energy, 1 (1957) 16
9. Hall, J. et al. Proc. 28th ICRC, Tsukuba, (2003) 2851
10. Bradbury, S.M. & Rose, H.J. Nucl. Instr. & Meth. A, 481 (2002) 521
11. Buckley, J.H. et al. Proc. 28th ICRC, Tsukuba, (2003) 2827
12. Holder, J. et al. Proc. 29th ICRC, Pune, 5, (2005) 383.
13. Maier, G. et al. Proc. 29th ICRC, Pune, 5, (2005) 395.

III Galactic Variable Sources

GALACTIC VARIABLE SKY WITH EGRET AND GLAST

S. W. DIGEL

Stanford Linear Accelerator Center
2575 Sand Hill Road Menlo Park, CA 904025, USA
** E-mail: digel@slac.stanford.edu*

The characteristics of the largely-unidentified Galactic sources of gamma rays that were detected by EGRET are reviewed. Proposed source populations that may have the correct spatial, spectral, luminosity, and variability properties to be the origins of the EGRET sources are also presented. Finally, the prospects for studying Galactic gamma-ray sources with the GLAST LAT are reviewed.

Keywords: Milky Way; gamma-ray

1. Introduction

The Energetic Gamma-Ray Experiment Telescope (EGRET) on the *Compton* Gamma-Ray Observatory provided a tremendous advance in high-energy gamma-ray astronomy. The number of known gamma-ray sources was increased from ~20 to ~300 [1], and although the positional determinations generally were not accurate enough to permit identifications of counterparts, overall average properties of the underlying source populations could be inferred.

The distribution of the sources on the sky suggests that a large fraction of the EGRET sources are clearly within the Milky Way. Few sources were observed deeply enough or often enough for variability to be detected with high confidence but overall many of the Galactic sources detected by EGRET are probably variable.

Several plausible candidate populations for Galactic sources have been described in the literature. All are related in one way or another to massive star formation or remnants of massive stars.

This paper provides a summary of the properties of the EGRET sources as well as of the likely underlying source populations. The potential for GLAST to advance the study of variable Galactic sources and source populations is also illustrated.

2. Properties of EGRET sources

The 3EG catalog has 271 sources, the majority of which were unidentified at the time of the production of the catalog and a few of which were known to be either transient (a solar flare) or probably spurious (e.g., point sources near the Vela pulsar that pulse in phase with it) [1] . The distribution of the sources on the sky suggested perhaps 3 populations: isotropic (presumably extragalactic), Galactic disk, and Galactic halo.

The isotropic component included the ~70 gamma-ray blazars that had been identified at the time, and since then likely blazar counterparts have been proposed for most of the high-latitude EGRET sources [2,3].

The Galactic halo sources, which were also characterized as 'faint' and 'steady' (see below), were initially proposed to be correlated with Gould's Belt, the relatively nearby star-forming regions that ring the sky [4,5] and not actually associated with the halo of the Milky Way. In this model the sources were considered most likely to be gamma-ray pulsars, the only identified Galactic population of gamma-ray sources. Deep searches of the error boxes for these sources for radio pulsars were subsequently undertaken, essentially unsuccessfully (e.g., Ref. 6).

More recently, the existence of these halo sources as point sources of gamma rays has been called into question based on a re-evaluation of the diffuse gamma-ray emission of the Milky Way [7]. The diffuse emission is a relatively bright background against which gamma-ray point sources must be detected. Owing to the limited statistics and angular resolution of EGRET data, the emission is typically modeled rather than inferred directly from the observations. Models for the diffuse emission based on interactions of cosmic rays with interstellar gas (as traced by observations of the 21-cm line of H I and the 2.6-mm line of CO, as a surrogate for H_2) and the interstellar radiation field have been very successful in describing the large-scale features of the gamma-ray sky. At intermediate latitudes, however, recent studies of infrared emission from interstellar dust suggest that some interstellar gas not previously accounted for (presumably H_2 that is not being traced by CO) is present at signficant levels [7]. Indeed a survey of 'infrared excess' clouds finds a large fraction to be undetectable in CO [8]. The reanalysis by Grenier & Casandjian of the EGRET data with a model of diffuse emission that includes infrared excess clouds suggests that the great majority of the faint, steady halo sources were misidentified diffuse emission [9].

The Galactic disk component of the EGRET sources is relatively narrowly distributed around the Galactic equator. The scale height ($< 2°$) and distribution in longitude suggest a characteristic distance of 1-6 kpc [10], i.e., not even as great as the distance to the Galactic center. The corresponding luminosities are $(1 - 15) \times 10^{35}$ erg s^{-1}, or $\sim$100 $L_{\odot}$ [10]. These sources and their potential origins are subject of the next sections.

2.1. *Variability of EGRET sources*

The limited sensitivity of EGRET relative to the characteristic fluxes of Galactic sources made short-term (scale of days) variability difficult to measure. A systematic search by Wallace et al. [11] found only 2 examples of possibly-significant day-scale flares; notably 3EG J0241+6103 (a possible counterpart to the microquasar LS I $+61^{\circ}$ 303) apparently doubled its flux for a $\sim$2-day interval.

On somewhat longer time scales of 1-2 weeks, the duration of a typical EGRET viewing period, the fluxes and upper limits published in Ref. 1 for each source can be used to estimate variability. Several authors have analyzed variability of the EGRET sources with this method, e.g., Refs. 12–14. As Reimer [15] points out, the resulting measures of variability were not always consistent between approaches, which differed notably in how upper limits were treated.

Typically, the variability index for a given source in the 3EG catalog is not very prescriptive about whether the source is variable. When the sources are considered as classes, then some plausible conclusions may be reached. For example, Nolan et al. [14] find that the identified blazars are signficantly variable sources and the pulsars (averaged over many spin periods of course) are not. Nolan et al. find that among the low-latitude sources, a large fraction of those toward the inner Galaxy (with $|l| < 55^{\circ}$) are significantly variable, with estimated dispersion of measured fluxes greater than 70% of the average flux. This distribution is consistent with the general direction of massive star-forming regions in the inner spiral arms of the Galaxy.

For the variable EGRET sources, Roberts et al. point out that time scales of variability shorter than months imply angular sizes of less than $1'$, and that particle acceleration happens on this time scale as well [16].

3. Candidates for variable Galactic sources

In this section, plausible (and published) classes of variable Galactic sources are described; all can produce sources of the required luminosities. In this context, rotation-powered pulsars are not considered to be variable sources; see the contribution by M. Razzano elsewhere in this volume.

3.1. *X-ray binaries: microquasars/microblazars*

X-ray binaries are close binary systems of a pulsar and a star, so-named because they are strong and variable X-ray sources. In high-mass X-ray binaries (HMXRB), the companion is an early-type star, with strong stellar winds. In low-mass X-ray binaries (LMXRB) the companion is less massive ($< 2.5\ M_\odot$) and accretion onto the pulsar is via Roche lobe overflow.

Several hundred XRBs are known in the Milky Way [17,18]. More have recently been discovered in hard X-ray observations with Integral [19]; these 'cocooned' binary systems are completely obscured by intervening interstellar gas at softer X-ray energies.

The distributions of HMXRB and LMXRB on the sky are significantly different [20]. The HMXRB are more tightly confined to the Galactic equator; LMXRB have a broader distribution in Galactic latitude and are not as tightly correlated with star-forming regions. Just as EGRET did not detect all of the GeV sources of gamma rays in the Milky Way, the known XRBs are flux limited; the Grimm et al. [20] sample includes all XRBs that reached a flux of 5 mCrab for the All-Sky Monitor of RXTE during the first 5 years of the mission. No complete census exists of XRBs; the gamma-ray manifestations detected by the LAT may well be a 'deeper' sample.

Microquasars are XRBs with relativistic radio jets, and the name 'microquasar' is intended to suggest an analogy with active galaxies, in which accretion onto a supermassive black hole powers jets on a much larger scale. (Carrying this analogy further, a microblazar would be a microquasar with a jet aligned with the line of sight to the earth; see below.) The inferred population of relativistic electrons in the jet of a microquasar can produce gamma rays through scattering of synchrotron photons in the jet (synchrotron self-Compton scattering) or perhaps more likely inverse Compton scattering of ultraviolet photons from the companion star.

The protypical microquasar is LS 5039, a HMXRB in which the companion is an O7 star. (See Sect. 3.4 for an alternative interpretation of the source.) The microquasar identification and association with EGRET source 3EG J1824-1514 were proposed by Paredes et al. [21]. As Paredes et al. [22] point out LS 5039 should be intrinsically variable in high-energy gamma rays owing to the high eccentricity ($e = 0.41$) of the orbit, which causes the accretion rate onto the compact object and the intensity of the stellar radiation field in its vicinity to vary.

Variability at the 4.1 d period of LS 5039 was not seen in the EGRET data for 3EG J1824-1514, although the limited coverage and sensitivity of the EGRET observations of this source are not particularly constraining and periodic variations are also superimposed on variability of the jets. Variations of the jets are not periodic, but in general for microquasars the transition to a high (X-ray) luminosity state seems to be associated with launching relativistic ($\Gamma > 2$) [23].

In addition, in microquasars, absorption effects in the atmosphere of the companion star, or in its wind, or even eclipses by the companion could also introduce modulation of the gamma-ray emission at the orbital period.

For microblazars, relativistic boosting is another source of intrinsic variability. Kaufman Bernado et al. [24] point out that the accretion disk (and jet) of microquasar/blazar will precess if the orbit of the companion star is not coplanar with the disk.

Paredes [25] published a list of 15 known or suspected microquasars in the Milky Way. Only 2 or 3 of these (LS I +61° 303, LS 5039, and Cygnus X-3) are potential counterparts to EGRET sources, and each of these is a HMXRB. LMXRBs are disadvantaged relative to HMXRBs as gamma-ray sources owing to the greatly decreased intensity of the UV radiation fields of the stellar companions. Grenier, Kaufman Bernado, & Romero [7] considered LMXRBs as potentially forming a population of variable, intermediate latitude gamma-ray sources. Their conclusion was that inverse Compton scattering of radiation external to the jet would not be sufficient for producing EGRET gamma-ray sources, although synchrotron self-Compton emission in the jets together with microblazar alignment of the jets might be feasible.

3.2. *Plerions: pulsar wind nebulae*

Plerions, also known as pulsar-powered nebulae or pulsar-wind nebulae, are 'filled-center' supernova remnants (SNR). A few dozen are known in the Milky Way; the Crab is the prototypical plerion but their properties vary widely owing to differences in, e.g., age of the SNR. A plerion can be identified as such via imaging X-ray observations without finding the pulsar driving the nebula. Searches of the error boxes of EGRET sources have yielded several plerions that are prospective high-energy gamma-ray sources (e.g., Refs. 27,28).

The mechanism for producing high-energy gamma-ray emission is likely to be synchrotron emission from electrons accelerated in the shock between the pulsar wind and interstellar medium [29]. The time scale for cooling via IC scattering is too long to be consistent with day-scale variability. Interaction with nearby molecular clouds, via Bremsstrahlung scattering, also can be ruled out owing to variability considerations. Roberts [29] reports that the 'best' candidate gamma-ray sources among pulsar wind nebulae are those for which the nebulae are ram pressure confined, meaning for which the pulsars have large space velocities. The reasoning is that compression and amplification of the magnetic field in the bow shock increases the maximum energy that electrons can achieve.

3.3. *Binary plerions*

As Dubus has pointed out, binary systems of plerions with early type stars may also be sources of high-energy gamma rays [30]. Actually, PSR B1259-63, a millisecond pulsar with a B0Ve companion in a 3.4-yr orbit has been detected at periastron by the H.E.S.S. ground-based gamma-ray telescope at energies >100 GeV [31]. Dubus pointed out that in LS 5039 and LS I+61°303 the resolved radio structures had not yet been demonstrated to be relativistic jets, and the masses of the compact objects in these sysetms were not known well enough to know whether they were not neutron stars. In the mechanism proposed by Dubus, the electrons accelerated in the pulsar wind nebula produce gamma rays by inverse Compton scattering on the optical-ultraviolet radiation field of the companion star. Binary plerion systems would necessarily be periodic gamma-ray emitters.

3.4. *Isolated black holes*

Punsley describes how rapidly-rotating, charged black holes could produce jets as a product of gravitohydrodynamic coupling of the magnetosphere to the rotating black hole [32]. The particles in the jet could emit high-energy gamma rays via inverse Compton scattering or possibly synchroton self-Compton scattering. In these systems, unlike rotation-powered pulsars, the magnetic field would be aligned with the rotation axis, so they would not pulsate. Variability on short time scales would be possible, e.g., from wobbling of the jet or a change of the injected spectrum of electrons.

4. Prospects for studying variable Galactic sources with the LAT

The Large Area Telescope (LAT) is the principal instrument on the GLAST mission, under development for launch in late 2007; the instrument and its performance are described in Ref. 33. For the present topic, the most relevant aspects are the large effective area (>8000 cm^2), very large field of view (>2 sr), and the frequent, uniform coverage of the sky that the

LAT will achieve. The LAT has no consumables, unlike EGRET, and the systematic uncertainties in flux measurements should be much less.

All of these considerations bode well for great advances in the study of variable Galactic gamma-ray sources with the LAT. The potential is illustrated in Fig. 1 via a simple simulation of a periodic source with properties like those expected for LS I +61° 303. This relatively-bright source against the relatively faint diffuse background in the outer Galaxy is probably a best case, but the example illustrates what will be possible for such a source in only weeks of surveying the sky; the design life of the mission is 5 years and the goal for operations is 10 years. The LAT certainly should 'see' much deeper into the Milky Way than EGRET was able to.

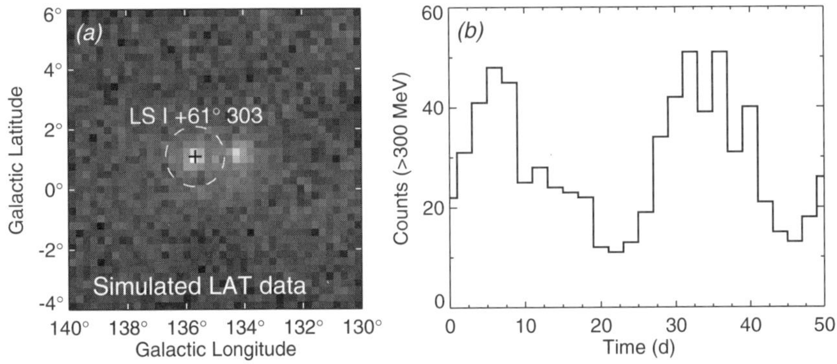

Fig. 1. Simulated LAT observation of the region around LS I +61° 303. The simulation is for the equivalent exposure of a 55-day sky survey. (*a*) Distribution of gamma rays with energies >300 MeV. The simulation includes a realistic model of the diffuse and point-source emission of the sky, and LS I +61° 303 has the flux of its potential counterpart 3EG source and a 26-day period with fairly strong modulation. (*b*) Counts per day for the region within the dashed circle (radius 1°).

5. Conclusions

Sources of gamma rays in the Milky Way are numerous and the evidence from EGRET is that many low-latitude sources (i.e., associated with Population I objects) are variable on time scales of days to months. Good candidates exist for variable Galactic sources - pulsar wind nebular, XRB (microquasars), or both (plerion binary systems). The variable Galactic gamma-ray sources will be much better characterized, more numerous, and it is to be hoped better understood, when GLAST is operational.

References

1. R. C. Hartman et al., *Astrophys. J. Suppl. Ser.*, *123*, 79 (1999).
2. D. Sowards-Emmerd, R. W. Romani, & P. F. Michelson, *Astrophys. J.*, **590**, 102 (2003).
3. D. Sowards-Emmerd et al., *Astrophys. J.*, **609**, 564 (2004).
4. N. Gehrels et al., *Nature*, **404**, 363 (2000).
5. I. A. Grenier, *Astron. & Astrophys.*, **364**, L93 (2000).
6. F. Crawford et al., *Astrophys. J.*, in press (2006).
7. I. A. Grenier, J.-M. Casandjian, & R. Terrier, *Science*, **307**, 1292 (2005).
8. T. Onishi et al., *Pub. Astron. Soc. Japan*, *53*, 1017 (2001).
9. I. A. Grenier & J.-M. Casandjian, in preparation (2006).
10. R. Mukherjee et al., *Astrophys. J.*, **441**, L61 (1995).
11. P. M. Wallace et al., *Astrophys. J.*, **540**, 184 (2000).
12. M. A. McLaughlin et al., *Astrophys. J.*, **473**, 763 (1996).
13. D. F. Torres et al., *Astron. & Astrophys.*, **370**, 468 (2001).
14. P. L. Nolan, W. F. Tompkins, I. A. Grenier, & P. F. Michelson, *Astrophys. J.*, **597**, 615 (2003).
15. O. Reimer, in *Nature of Unidentified Galactic High-Energy Gamma-Ray sources*, (Kluwer, Dordrecht, 2001).
16. M. S. E. Roberts, B. M. Gaensler, & R. W. Romani, in *Neutron Stars in Supernova Remnants*, 213 (ASP, San Francisco, 2002).
17. Q. Z. Liu, J. van Paradijs, & E. P. J. van den Heuvel, *Astron. & Astrophys. Suppl.*, **147**, 25 (2000).
18. Q. Z. Liu, J. van Paradijs, & E. P. J. van den Heuvel, *Astron. & Astrophys. Suppl.*, **368**, 1021 (2001).
19. R. Walter et al., *Astron. & Astrophys.*, **453**, 133 (2006).
20. H.-J. Grimm, M. Gilfanov, & R. Sunyaev, *Astron. & Astrophys.*, **391**, 923 (2002).
21. J. P. Paredes et al., *Science*, 288, 2340 (2000).
22. J. P. Paredes et al., *Astron. & Astrophys.*, **393**, L99 (2003).
23. R. Fender & T. Maccarone, in *Cosmic Gamma-Ray Sources*, 205 (Kluwer, Dordrecht, 2004).
24. M. M. Kaufman Bernado, G. E. Romero, & I. F. Mirabel, *Astron. & Astrophys.*, **385**, L10 (2002).
25. J. M. Paredes, *Chin. J. Astron. & Astrophys. Suppl.*, **5**, 121 (2005).
26. I. A. Grenier, M. M. Kaufman Bernado, & G. E. Romero, *Astrophys. & Space Sci.*, **297**, 109 (2005). ref for number of Plerions known in MW
27. M. S. E. Roberts et al., *Astrophys. J.*, **515**, 712 (1999).
28. J. P. Halpern et al., *Astrophys. J.*, **552**, L125 (2001).
29. M. S. E. Roberts et al., *Astrophys. & Space Sci.*, **297**, 93 (2005).
30. G. Dubus, *Astron. & Astrophys.*, **456**, 801 (2006).
31. F. Aharonian et al., *Astron. & Astrophys.*, **442**, 1 (2005).

32. B. Punsley, *Astrophys. J.*, **498**, 640 (1998).
33. J. E. McEnery, I. V. Moskalenko, & J. F. Ormes, in *Cosmic Gamma-Ray Sources*, 361 (Kluwer, Dordrecht, 2004).

Galactic Variable Sources Observed with H.E.S.S.

Nukri Komin
for the H.E.S.S. collaboration
CNRS/IN2P3/LPTA Montpellier
Place Eugene Bataillon, Batiment 13
F-34095 Montpellier Cedex 5
E-mail: komin@lpta.in2p3.fr

H.E.S.S., a system of imaging Cherenkov telescopes, is dedicated to the observation of TeV gamma-rays. Within the first years of operation a number of objects were detected, most of these objects were previously not known to be TeV emitters. The observed TeV emission is crucial for the understanding of particle acceleration in the sources.

Here I will review the results obtained on Galactic sources expected to show variable emission. Variable emission was detected with high significance from the binary systems PSR B1259−63 / SS 2883 and LS 5039. The emission of the latter object appears to be periodic in accordance with the orbit. No pulsed emission from pulsars was detected so far. For three pulsars (PSR B0531+21, PSR B0833−45, PSR B1706−44) upper limits on the TeV emission at the EGRET pulse phases were derived.

Keywords: H.E.S.S.; Galactic sources; Variability.

1. The H.E.S.S. Detector

H.E.S.S. (High Energy Stereoscopic System) is a system of four imaging Cherenkov telescopes dedicated to the observation of gamma-ray emission in the energy range between several hundreds of GeV up to several tens of TeV. It is located in Namibia in the south of Africa. Each of the telescopes is equipped with a tessellated mirror with a surface of 107 m^2 and a phototube camera comprising 960 pixels. The cameras have a field of view with a diameter of 5°. Stereoscopic observations of air showers allows the reconstruction of the direction of the individual photons with an accuracy better than 0.1°. The energy of each photon can be determined with a resolution of about 15% (For details see Reference 1). The unprecedented sensitivity of H.E.S.S. and its location in the southern hemisphere makes it an ideal instrument for the observation of Galactic sources. Within its first three

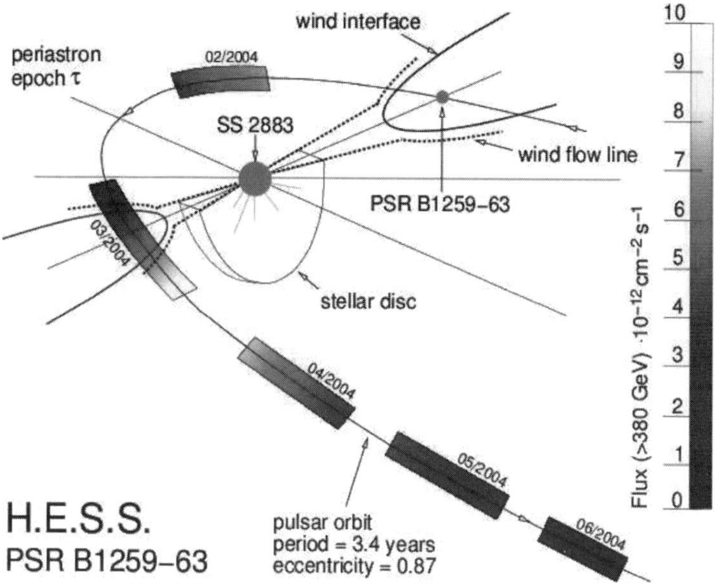

Fig. 1. Sketch of the orbit of PSR B1259−63 (picture taken from Refs. 2,3). The intensity scale denotes the gamma-ray flux observed by H.E.S.S. at the different orbital phases.

years of observation about 35 sources have been detected.

Each event observed with H.E.S.S. is recorded with a GPS time-stamp allowing the search for variability and periodicity. Periodic emission of Galactic sources can be expected in a large range of time-scales, from years in binary systems down to milliseconds for pulsars.

2. PSR B1259−63 / SS 2883

The binary system PSR B1259−63 / SS 2883 consists of a pulsar orbiting a massive, luminous star. The orbit is highly eccentric with a period of 3.4 years and an inclination of 35°. A sketch of the orbit is shown in Fig. 1. During periastron the distance between the pulsar and the star is about 10^{13} cm.

The pulsar PSR B1259−63 was detected and observed in radio and has a period of 48 ms. The radio emission is eclipsed by the star during periastron. The companion B2e-type star SS 2883 has a mass of about ten solar masses ($10\,M_\odot$) and a mass outflow forming an equatorial disc. This disc is crossed by the pulsar before and after periastron. The star's dense photon field

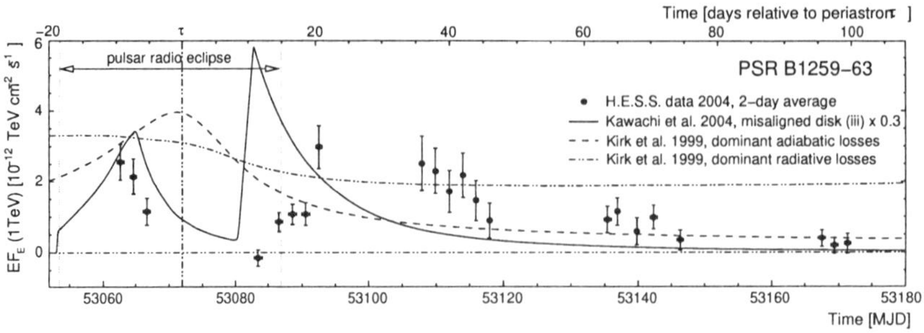

Fig. 2. Light curve of the TeV emission of PSR B1259−63 around periastron (τ, dotted line) [3]. The passage of the equatorial disc is marked with dashed lines. Models of an electronic emission scenario [4] and a hadronic emission scenario [5] are indicated.

provides an ideal target for inverse Compton scattering of the electrons and positrons accelerated in the vicinity of the pulsar.

The binary system was observed with the H.E.S.S. telescopes around periastron in 2004 [3]. Gamma-ray emission of the system was detected with a significance of $13.8\,\sigma$. The size of the emission region is consistent with a point-like source. Thus, given the H.E.S.S. point-spread function and assuming a distance of the system from the observer of 1.5 kpc, the emission region is smaller than 0.24 pc.

The gamma-ray emission was found to be clearly variable. PSR B1259−63 / SS 2883 is thus the first Galactic source observed with H.E.S.S. showing variable emission. Figure 2 shows the light curve of the gamma-ray emission. It shows an decrease of the emission before the periastron and an increase after periastron, during the second disc passage. Full moon prevented observations of the system during periastron. Later on, with increasing distance between the pulsar and companion star, the gamma-ray emission decreases and is not detectable about 100 days after periastron. No variations of the spectral index have been found. The integrated photon flux at the different orbital positions of the pulsars are shown in Fig. 1.

Possible scenarios for the emission are inverse Compton emission of electrons and positrons or emission of decaying neutral pions produced in hadronic interactions. Inverse Compton emission of electrons off the stellar photon field should reach its maximum in the dense stellar photon field near periastron. Models for radiative and adiabatic losses of electrons [4] are indicated in Fig. 2. On the other hand, the equatorial disc provides target material for hadronic interactions. A hadronic model of the emission [5] (solid line in Fig. 2) predicts a peak of gamma-ray emission before periastron and a steep rise of the emission before the second disc passage. Given the present H.E.S.S. data neither of the models is favoured to explain the observed light curve.

Continued observations of the source will sample the whole orbit of the system. The present data suggest that the long term emission depends on the orbital period of the system showing its maximum near periastron. The emission shows variability on time-scales of days. The mechanism leading to the gamma-ray emission remains unclear; the equatorial disc of the star may play a crucial role. Observations of the next periastron with H.E.S.S. and other gamma-ray experiments may clarify the nature of the gamma-ray emission mechanisms.

3. The X-ray binary LS 5039

X-ray binaries are systems of compact massive objects (neutron stars or black holes) orbiting a companion star. They may have radio jets of particle outflow, similar to Active Galactic Nuclei (AGN).

The system LS 5039 consists of a compact object with a mass of $\sim 1.38\,M_\odot$ orbiting an O-type star with a mass of $\sim 20\,M_\odot$ and a radius of $\sim 9.3\,R_\odot$. Whether the compact object is a neutron star or a black hole is up to now unclear. The orbit has a low eccentricity of 0.35, an inclination of 25° and a period of 3.9 d [6].

Gamma-ray emission from the direction of LS 5039 was detected in the data of the first Galactic plane survey conducted with H.E.S.S. [7]. An excess of 250 gamma-rays with a significance of $8\,\sigma$ was detected [8]. The observed excess is a point-like source; its position is consistent with radio data obtained on LS 5039. Within the system particles can be accelerated in an AGN-like jet or in a shock front formed between a possible pulsar wind nebula and the stellar wind. Inverse Compton scattering of accelerated electrons and positrons [9] or pion production by the interaction of accelerated hadrons [10] could lead to TeV production. If the TeV photons are produced in interactions with the stellar photon field one would expect a maximum of the emission near periastron where the density of the stellar photon field seen by the pulsar reaches its maximum. Absorption of the gamma-rays in the stellar photon sphere will further modulate the light-curve, shifting the maximum of the emission towards the inferior conjunction [11].

In order to test for periodicity of the gamma-ray emission further observations of the source have been carried out in 2005 [12]. The analysis of the data with a Lomb-Scargle-Test [13] shows a periodicity of 3.9078 ± 0.0015

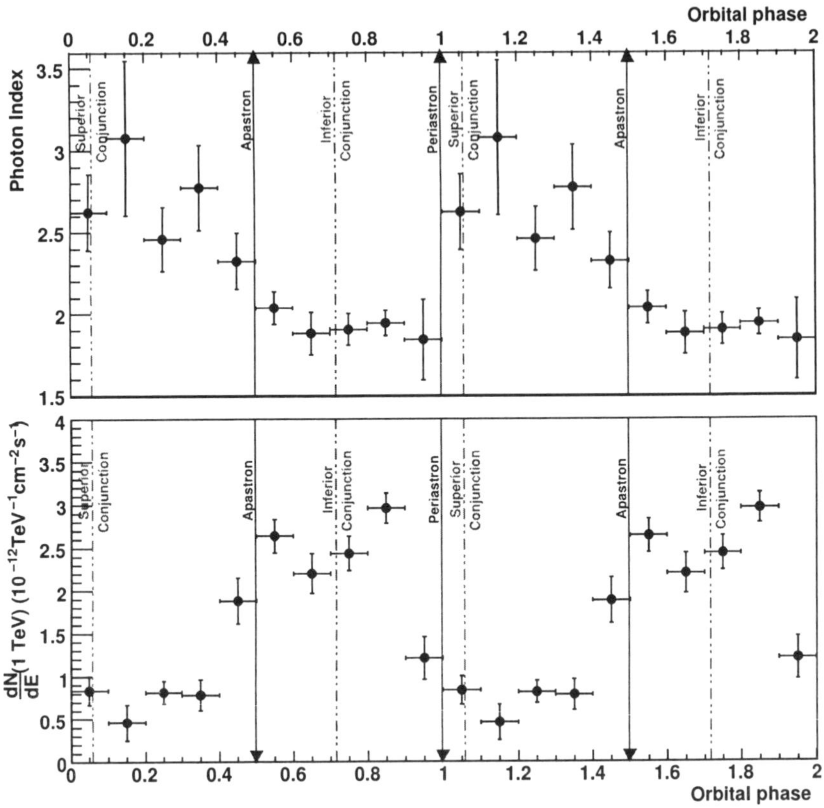

Fig. 3. Normalisation (bottom) and photon index of the gamma-ray emission of LS 5039 in dependence on the orbital phase [12]. For clarity, two phases are shown.

days, consistent with the period derived in Ref. 6. The gamma-ray excess was divided in bins of orbital phase and fitted by a power law. Figure 3 shows the normalisation of the power law at 1 TeV (bottom panel) and the photon index (top panel). It can be seen that the maximum of the emission is aligned with the inferior conjunction, the minimum is aligned with the superior conjunction. The data also show a significant change of the spectral index with the orbital period. A hard gamma-ray spectrum (photon index $1.85 \pm 0.06_{\rm stat} \pm 0.1_{\rm syst}$) was observed during inferior conjunction, while the spectrum at superior conjunction is relatively steep (photon index $2.53 \pm 0.07_{\rm stat} \pm 0.1_{\rm syst}$). LS 5039 is thus the first Galactic object showing temporal variations in the spectral shape of the TeV gamma-ray emission.

Finding the maximum and minimum of the emission at the conjunctions is consistent with absorption of the gamma rays in the stellar photosphere [11]. A detailed interpretation of the observed light curve can be found in Ref. 12: The emission region cannot be significantly larger than the orbit, otherwise the modulation would be smeared out over all phases. The variations of the spectral index can be explained by a change of the maximum electron energy. During apastron (close to inferior conjunction for LS 5039) the maximum electron energy is higher, thus allowing a harder photon spectrum. Strong synchrotron losses near periastron could lead to a steeper photon spectrum. The detection of gamma rays at superior conjunction is not consistent with absorption models. Thus, further processes may play a role in the TeV emission of X-ray binaries.

The gamma-ray emission of LS 5039 shows periodic gamma-ray emission and a change of the photon index in accordance with the orbital period of the system. It provides insights in the acceleration and absorption mechanisms in X-ray binaries and establishes X-ray binaries as a new class of TeV gamma-ray emitters.

4. Pulsars

Pulsars are neutron stars with a high magnetic field. They are fast rotating with periods of milliseconds to seconds. Pulsed emission from pulsars was first discovered in radio, and since then detected at optical wavelengths, in X-rays and in gamma rays. Most of the pulsars with gamma-ray emission show a cut-off of the spectrum at an energy of several GeV.

Two different scenarios place the emission region near the magnetic poles of the pulsar (polar cap model) or near the null surface in the outer magnetosphere (outer gap model) (for a review see e.g. Ref. 14). These models predict different cut-offs of the spectrum at GeV energies, an energy band which is not observable by current instruments. In addition, the outer gap model predicts a peak of inverse Compton emission at TeV energies (see e.g. Ref. 15). Detection of pulsed TeV gamma-ray emission would thus favour the outer gap model.

Here I will review the results of three pulsars observed with H.E.S.S. The Crab nebula is one of the brightest gamma-ray sources. It consists of the pulsar PSR B0531+21 and the surrounding pulsar wind nebula (PWN). The gamma-ray emission detected by H.E.S.S. is a point-like source positionally coincident with the pulsar and the PWN [1]. The spectral shape is compatible with synchrotron-self Compton emission of accelerated electrons. Thus, the emission is associated with the PWN.

The Vela pulsar (PSR B0833−45) and the PWN Vela X are located in

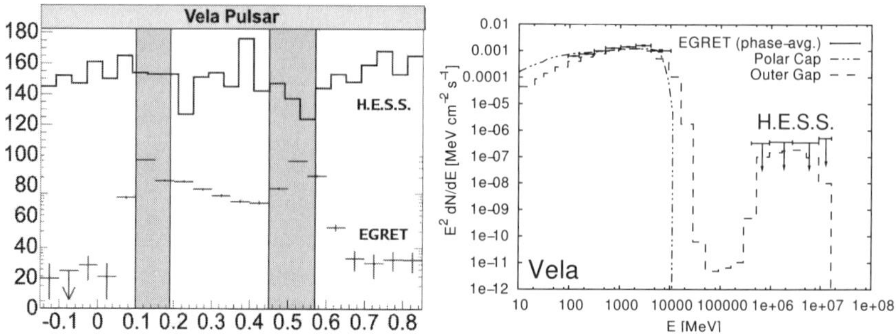

Fig. 4. Search for pulsed emission from the Vela pulsar (PSR B0833−45). The left panel shows the H.E.S.S. and EGRET data in a phasogram. The right panel shows the H.E.S.S. upper flux limits compared to the different emission models. The pictures are taken from Ref. 19.

the Vela supernova remnant. Gamma-ray emission of extended morphology and offset from the pulsar was detected with H.E.S.S. [16]. Based on the morphological match with X-ray data the emission was found to be originating from the PWN. No emission was found from the direction of the pulsar itself [17].

PSR B1706−44 is a young pulsar detected in radio, X-rays and GeV gamma rays. An extended synchrotron nebula surrounding the pulsar was observed in radio and in X-rays, suggesting the existence of a PWN. This object was observed with the H.E.S.S. telescopes but no gamma-ray emission has been found from neither the pulsar nor the PWN [17,18].

For these three pulsars a search for pulsed emission was performed [19]. The arrival times of the detected events, transformed into the solar system's barycentre, were folded into a phasogram using radio ephemerides from Jodrell Bank[a] and the Australian Pulsar Timing Archive[b]. The phasogram of PSR B0833−45 for H.E.S.S. and EGRET observations are shown in the left panel of Fig. 4. The curve of the EGRET data clearly shows a double peak. The H.E.S.S. phasogram is featureless and does not show any hint for pulsed emission.

For further analysis the phase range of the EGRET detected pulses were chosen as on-pulse or test regions (gray areas in Fig. 4), the remaining regions were used as off-pulse region for background estimation. Based on these regions upper limit of the pulsed photon flux at the position of the EGRET pulses were calculated and are summarised in Table 1.

[a]http://www.jb.man.ac.uk/ pulsar/crab.html
[b]http://www.atnf.csiro.au/research/pulsar/archive/

Table 1. Upper limits (99.9% confidence level) for the integrated flux of pulsed emission from pulsars.

Pulsar	energy threshold	integrated flux above threshold
PSR B0531+21	560 GeV	$6.30 \times 10^{-12}\,\mathrm{cm^{-2}s^{-1}}$
PSR B0833−45	400 GeV	$0.50 \times 10^{-12}\,\mathrm{cm^{-2}s^{-1}}$
PSR B1706−44	500 GeV	$1.38 \times 10^{-12}\,\mathrm{cm^{-2}s^{-1}}$

The obtained upper limits can be used to test the outer gap model of pulsar emission. The right panel of Fig. 4 shows the H.E.S.S. upper limits in comparison with the predictions by the different models. The upper limits are at the level of the predictions for the inverse Compton component of the outer gap model. Further observations will be needed to increase the statistics and to put constraints on the outer gap model.

In the Galactic plane survey conducted with H.E.S.S. [7,20] data on more pulsars were obtained. Statistical tests will be applied to search for pulsed emission independent of EGRET data. In order to reach the interesting energy region near the cut-off, the gamma-ray selection cuts will be optimised for low energies. Finally, this interesting energy region will be available for future instruments like GLAST and H.E.S.S. II.

5. Summary

Within the first years of operation of H.E.S.S. a large number of sources were detected. Most of the Galactic sources are of extended morphology, like pulsar wind nebulae, and are not expected to show variability or periodic emission.

Two Galactic sources, the binary systems PSR B1259−63 / SS 2883 and LS 5039, show variability. Their emission is found to be modulated with their orbital period. The observational results provide insight into the particle acceleration mechanisms in these systems. For PSR B1259−63 / SS 2883 no final conclusion on the gamma-ray production can be drawn yet and further observations of the upcoming periastron are necessary.

Up to now, no pulsed emission from pulsars have been detected with H.E.S.S. Observation of pulsed gamma rays at the high-energy end of the pulsed emission will test the emission models of pulsars and will be a major task of future gamma-ray instruments.

References

1. Aharonian, F.A. et al. (H.E.S.S. collaboration), *A&A* **457**, 899(Octtober

2006).
2. S. Johnston, R. N. Manchester, D. McConnell and D. Campbell-Wilson, *MNRAS* **302**, 277(January 1999).
3. Aharonian, F.A. et al. (H.E.S.S. collaboration), *A&A* **442**, 1(Octtober 2005).
4. J. G. Kirk, L. Ball and O. Skjaeraasen, *Astroparticle Physics* **10**, 31(January 1999).
5. A. Kawachi, T. Naito, J. R. Patterson et al., *ApJ* **607**, 949(June 2004).
6. J. Casares, M. Ribó, I. Ribas, J. M. Paredes, J. Martí and A. Herrero, *MNRAS* **364**, 899(December 2005).
7. Aharonian, F.A. et al. (H.E.S.S. collaboration), *Science* **307**, 1938(March 2005).
8. Aharonian, F.A. et al. (H.E.S.S. collaboration), *Science* **309**, 746(July 2005).
9. V. Bosch-Ramon and J. M. Paredes, *A&A* **417**, 1075(April 2004).
10. F. Aharonian, L. Anchordoqui, D. Khangulyan and T. Montaruli, *Journal of Physics Conference Series* **39**, 408(May 2006).
11. G. Dubus, *A&A* **451**, 9(May 2006).
12. Aharonian, F.A. et al. (H.E.S.S. collaboration), 3.9 day orbital modulation in the TeV -ray flux and spectrum from the X-ray binary LS 5039 A&A in press, astro-ph/0607192, (2006).
13. J. D. Scargle, *ApJ* **263**, 835(December 1982).
14. A. K. Harding, Emission From Rotation-Powered Pulsars and Magnetars, in *AIP Conf. Proc. 801: Astrophysical Sources of High Energy Particles and Radiation*, eds. T. Bulik, B. Rudak and G. MadejskiNovember 2005.
15. J. Takata, S. Shibata, K. Hirotani and H.-K. Chang, *MNRAS* **366**, 1310(March 2006).
16. Aharonian, F.A. et al. (H.E.S.S. collaboration), *A&A* **448**, L43(March 2006).
17. B. Khélifi, Nu. Komin, T. Lohse et al., Search for TeV Emission from the Direction of the Vela and PSR B1706-44 Pulsars with the H.E.S.S. Experiment, in *AIP Conf. Proc. 745: High Energy Gamma-Ray Astronomy*, eds. F. A. Aharonian, H. J. Völk and D. HornsFebruary 2005.
18. Aharonian, F.A. et al. (H.E.S.S. collaboration), *A&A* **432**, L9(March 2005).
19. F. Schmidt, F. Breitling, S. Gillessen et al., Search for Pulsed TeV Gamma-Ray Emission from Young Pulsars with H.E.S.S., in *AIP Conf. Proc. 745: High Energy Gamma-Ray Astronomy*, eds. F. A. Aharonian, H. J. Völk and D. HornsFebruary 2005.
20. Aharonian, F.A. et al. (H.E.S.S. collaboration), *ApJ* **636**, 777(January 2006).

Gamma-ray Pulsars in the GLAST era

M. RAZZANO

Universitá di Pisa and Istituto Nazionale di Fisica Nucleare Sezione di Pisa,
Largo B. Pontecorvo 3, 56127 Pisa, Italy
** E-mail: massimiliano.razzano@pi.infn.it*

Pulsars are useful tools for probing the law of physics under extreme conditions. Since pulsars emit radiation across the whole electromagnetic spectrum they are ideal targets for multiwavelength studies. Today we know a large number of radio pulsars but unfortunately our current knowledge of gamma-ray emitting pulsars is still lacking. Presently there are only seven high-confidence detections of gamma-ray pulsars, four of them detected by the instruments of the *Compton Gamma Ray Observatory*. The launch of GLAST and AGILE mission will lead to a major breakthrough in our understanding of gamma-ray pulsars. The general properties of the presently known gamma-ray pulsars are reviewed and the impact of the future gamma-ray mission on pulsar science is presented.

Keywords: Pulsars, gamma-ray, GLAST

1. Introduction

Pulsars are the brightest sources in the gamma-ray sky and act as unique laboratories for probing the physical processes under extreme physical environments They emit radiation in the whole electromagnetic spectrum and they are a perfect target for multi wavelength campaigns.[1]
Pulsar are rotating, highly-magnetized neutron stars emitting pulsed radiation because of rotation. A small fraction of the rotational energy is converted into radiation and the bulk of radiation is emitted in gamma-rays.
Pulsars were discovered in 1967 by Jocelyn Bell, a Ph.D. student working in Cambridge under the supervision of Antony Hewish on the study of planetary scintillation.[2] With the antenna designed for this study they discovered a radio source that eventually seemed to be located outside the Solar System. This source was named CP1919 and was studied with more sensitive radio telescope, opening a window on a new class of sources that were called Pulsars (from *Pulsating Radio Sources*). Today more than 1700

radio pulsars are known, while pulsars in other wavelengths are more difficult to detect, because of a combination of emission geometry and telescopes sensitivity. About 80 x-ray pulsars are known and only 7 gamma-ray pulsars. The current knowledge about pulsars is still far to be complete, and the study and discovery of new gamma-ray pulsars will provide a perfect tool for probing these extreme intriguing sources.

1.1. *The standard model*

In the so-called "standard model"[3] of pulsars the pulsed emission is explained as a consequence of the rapid rotation of a neutron star with an high magnetic field. Observed periods ranges over some order of magnitude, from about $\simeq$1.6 ms to $\simeq$ 8.5s.[4] In 1969 Goldreich and Julian solved the equation of an aligned rotator in vacuum,[5] showing that the environment around the pulsar must be filled by a plasma forming a *magnetosphere* corotating with the star. Since the magnetosphere is rotating at the same angular speed of the star, there is a critical radius, named radius of the *light cylinder* R_c, where the particles in the magnetosphere should corotate at the speed of light. The field lines at this distance would eventually swept out due to relativistic effects, creating some field lines that close at a distance larger than R_c (See Fig.1) Particles can be extracted from the

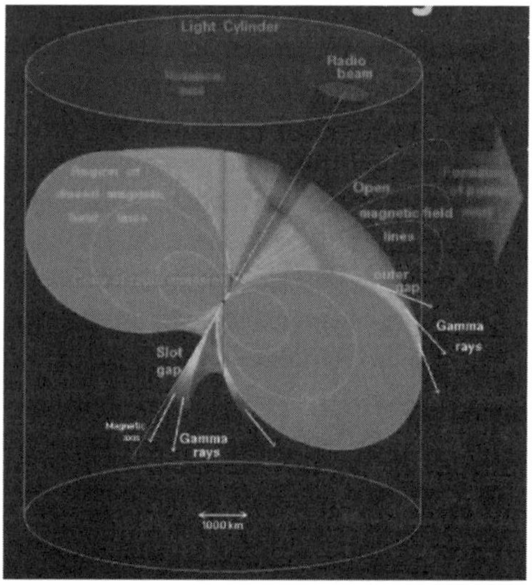

Fig. 1. A scheme of the standard model for pulsars

surface since the electric fields are several orders of magnitude greater than the gravitational force. They are accelerated along the magnetic field lines. The particles traveling along the open field lines can reach Lorentz factors of about $\gamma \simeq 10^7$, and eventually emit radiation. The radiation losses are then not only due to magnetic dipole radiation but mainly due to magnetospheric emission. The last closed field lines identified a region on the neutron star surface called *Polar Caps*, located around the magnetic poles of the star. The bulk of the rotational energy is converted into pulsar wind, while a minor fraction is converted into radiation.
According to the standard model the radio emission takes place above the magnetic poles and the rotation of the star produce a pulsed emission because the rotation axis and the magnetic axis are misaligned, as displayed in Fig.1. The emission pattern sweep the line of sight of the observers as in a lighthouse. The gamma-ray emission takes place along the open field lines, where particles can be accelerated up to very high energies.

2. Gamma-ray pulsars in the EGRET era

The three brightest point sources in the gamma-ray sky appeared to be the Vela (PSR B0833-45), Crab (PSR B0531+21) and Geminga (PSR J0633+1746), already from the observations of the *Small Astronomy Satellite* (SAS-2) and COS-B. Pulsation of Vela and Crab were detected by SAS-2 and extended by the COS B mission. A major step in understanding gamma-ray pulsars came with the launch of the *Compton Gamma-ray Observatory* (CGRO). CGRO carried four experiments (COMPTEL, OSSE, BATSE, EGRET) that increased to seven the number of currently known gamma-ray pulsars. EGRET was a pair conversion telescope that discovered the gamma-ray emission from PSR B1706-44,[6] PSR B1055-52,[7] PSR B1951+32[8] and detected the modulation of the gamma-rays from Geminga.[9] The BATSE telescope, mainly devoted to Gamma Ray Bursts, detected the pulsar PSR B1509-58,[10] that was observed also by COMPTEL up to 10 MeV. Some general facts can be derived by looking at the seven high-confidence gamma-ray pulsars in a multiwavelength context and by comparing them to the population of radio pulsars. The lightcurves look different at different energies, then we can infer that the emission mechanisms are the result of a combination of geometry and energy band. For most of them the lightcurve has two peaks. Not all have been observed at EGRET energies, e.g. PSR B1509-58 as been seen by COMPTEL up to 10 MeV. By looking at the multiwavelength spectra in Fig.2 it is possible to see that the maximum power output is in gamma-rays. There is a dis-

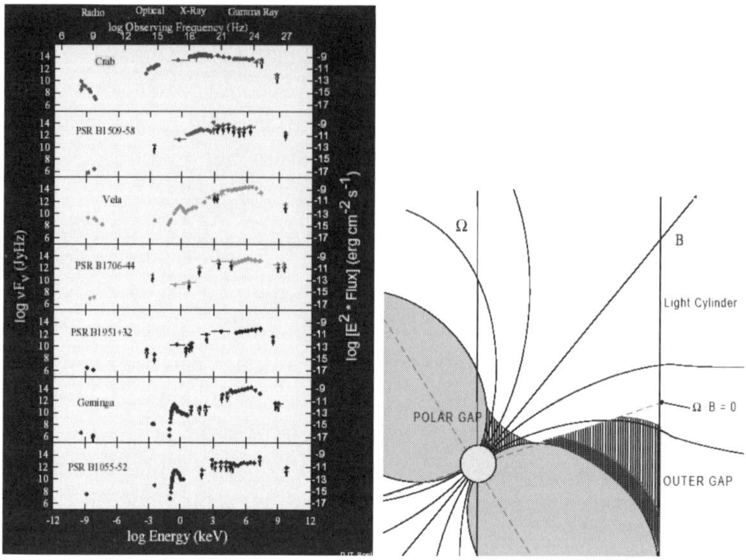

Fig. 2. Right: Multiwavelength spectra $\nu F\nu$ of the EGRET pulsars (from Thompson(2001)[1]). Left:A scheme of the Polar Cap and Outer Gap acceleration regions (from Harding (2001[11]).

tinction between the emission in radio (probably coming from a coherent process) and in gamma-rays, that seems to be due to single particle emission in incoherent processes. In case of Vela, Geminga and PSR B1055-52 a clear thermal component is visible, probably from the hot neutron star surface. Some pulsar show energy breaks at Gev energies, that can be related to the high magnetic fields at the neutron star surface. There is no detected pulsed emission at TeV energy, so that an energy cutoff is expected at energies greater than EGRET band. The three brightest gamma-ray pulsars (Vela, Geminga and Crab) show a phase-dependent spectrum with no simple phase-energy pattern, while for the others a phase-resolved spectroscopic study is not possible because of the low statistics. By looking at the $P - \dot{P}$ diagram it is visible that these pulsars are young objects with high surface magnetic fields and high open field line voltage. Although there is a concentration of millisecond radio pulsars, there's not yet an high-confidence detection of a gamma-ray millisecond pulsar, with exception for a low-confidence detection of PSR J0218+4232.[1] Among gamma-ray pulsars a particular mention is for Geminga, that lacks of a radio-counterparts. Geminga is the first representative of a potential new population of gamma-ray pulsars, often called as radio-quiet pulsars. Geminga was discovered as point sources by SAS-2 in the region of the Galactic Anticenter and an X-ray counterparts was found in the data of the Einstein satellite . A periodic signal of about 237 ms period was then found in the EGRET data and then retrieved in the data of COS B and SAS-2. In addition to these seven pulsars there are three additional low-confidence pulsars.[1] PSR B1046-58 may can be associated to 3EG J1048-5840 and has an associated X-ray source that is not pulsed. The second one is PSR B0656+14 is not coincident with any 3EG sources but has an optical and X-rays pulsed component. The third low-confidence pulsar is J0218+4232 and is of particular importance because is a millisecond pulsar, but the detection was complicated by the presence of a nearby blazar.[1]

3. GLAST and gamma-ray pulsars

According to the presently status of observations there are two main classes of models describing gamma-ray emission from pulsars. In the so-called Polar Cap models[12] the emission takes place at low altitude above Polar Caps, while in the Outer Gaps models[13,14] the gamma-ray emission comes from a region near the light cylinder. Recently a modification of the Polar Cap scenario has been proposed,[11] where emission takes place in a Slot Gap at high altitude above polar caps. A scheme of these acceleration region can be found in Fig.2.

These models make some distinct predictions on the gamma-ray emission from pulsars, then missions like GLAST[15] or AGILE will be able to distinguish between them. In particular the two models make different predictions on the high-energy cutoff, which is more soft in the Outer Gap models than in the Polar Cap models. The other important point that distinguish the Polar Cap from Outer Gap models is the ratio of radio-loud to radio-quiet pulsars.[11] The high energy coverage of GLAST will permit the study of the high-energy cutoff and the higher sensitivity will provide much statistics for blind searches and for the detection of new Geminga-like pulsars. An important contribution is expected also from the AGILE mission, which would cover a time window before the launch of GLAST. These future space missions will expand the number of known gamma-ray pulsars, providing new data for better understand the physics of these extremely fashinating objects.

References

1. D. J. Thompson, Gamma-ray Pulsars: Observations, in *American Institute of Physics Conference Series*, 2001.
2. A. Hewish, S. J. Bell, J. D. Pilkington, P. F. Scott and R. A. Collins, *Nature* **217**, 709 (1968).
3. P. A. Sturrock, *The Astrophysical Journal* **164**, 529(March 1971).
4. S. L. Shapiro and S. A. Teukolsky, *Black holes, white dwarfs, and neutron stars: The physics of compact objects* (Research supported by the National Science Foundation. New York, Wiley-Interscience, 1983, 663 p., 1983).
5. P. Goldreich and W. H. Julian, *The Astrophysical Journal* **157**, 869(August 1969).
6. D. J. Thompson, Z. Arzoumanian, D. L. Bertsch, K. T. S. Brazier, N. D'Amico, C. E. Fichtel, J. M. Fierro, R. C. Hartman, S. D. Hunter and S. Johnston, *Nature* **359**, 615(Octtober 1992).
7. J. M. Fierro, D. L. Bertsch, K. T. S. Brazier, J. Chiang, N. D'Amico, C. E. Fichtel, R. C. Hartman, S. D. Hunter, S. Johnston, G. Kanbach, V. M. Kaspi, D. A. Kniffen, Y. C. Lin, A. G. Lyne, R. N. Manchester, J. R. Mattox, H. A. Mayer-Hasselwander, P. F. Michelson, C. von Montigny, P. L. Nolan, E. Schneid and D. J. Thompson, *The Astrophysical Journal* **413**, L27(August 1993).
8. P. V. Ramanamurthy, D. L. Bertsch, B. L. Dingus, J. A. Esposito, J. M. Fierro, C. E. Fichtel, S. D. Hunter, G. Kanbach, D. A. Kniffen, Y. C. Lin, A. G. Lyne, J. R. Mattox, H. A. Mayer-Hasselwander, M. Merck, P. F. Michelson, C. von Montigny, R. Mukherjee, P. L. Nolan and D. J. Thompson, *Astrophysical Journal Letters* **447**, L109+(July 1995).
9. D. L. Bertsch, K. T. S. Brazier, C. E. Fichtel, R. C. Hartman, S. D. Hunter, G. Kanbach, D. A. Kniffen, P. W. Kwok, Y. C. Lin and J. R. Mattox, *Nature* **357**, 306(May 1992).
10. M. P. Ulmer, S. M. Matz, R. B. Wilson, M. J. Finger, K. S. Hagedon, D. A. Grabelsky, J. E. Grove, W. N. Johnson, R. L. Kinzer, J. D. Kurfess, W. R. Purcell, M. S. Strickman, V. M. Kaspi, S. Johnston, R. N. Manchester, A. G. Lyne and N. D'Amico, *The Astrophysical Journal* **417**, 738(November 1993).
11. A. K. Harding, Gamma-ray Pulsars: Models and Predictions, in *American Institute of Physics Conference Series*, 2001.
12. J. K. Daugherty and A. K. Harding, *AAPS* **120**, C107+(November 1996).
13. K. S. Cheng, C. Ho and M. Ruderman, *The Astrophysical Journal* **300**, 500(January 1986).
14. R. W. Romani and I.-A. Yadigaroglu, *The Astrophysical Journal* **438**, 314(January 1995).
15. N. Gehrels and P. Michelson, *Astroparticle Physics* **11**, 277(June 1999).

SOLVING THE RIDDLE OF UNIDENTIFIED HIGH-ENERGY γ-RAY SOURCES.

PATRIZIA A. CARAVEO
IASF-INAF, Via Bassini, 15 20133 Milano Italy
pat@iasf-milano.inaf.it

Unidentified objects dominate current catalogues of high-energy (MeV-to-GeV) sources. Solving the mystery of such unidentified γ-ray sources is a major challenge for astronomy. Different approaches towards source identification have been tried in the past, with limited success. The wealth of gamma-ray sources expected in the near future calls for a novel approach.

1. Unidentified γ-ray sources

Unveiling the nature of a vast number of unidentified sources is the most compelling problem facing today's high-energy (MeV-to-GeV) γ-ray astronomy. However, unidentified sources are not peculiar to high-energy γ-ray astronomy, they have been an ever-present phenomenon in astronomy. Indeed, every time a new astronomical window was opened, astronomers found sources they were not able to identify, i.e. to associate with previously known objects. This can happen either because such sources belong to a genuinely new (thus unknown) class or because their positions are not known accurately enough to allow for an unambiguous association between the newly found emitter and a known object. Thus, the lack of identification is frequently ascribed to poor angular resolution. Being unidentified, however, is a "temporary" status: sooner or later better tools will allow the source identification, i.e. either its classification within a given class of astronomical objects or its recognition as belonging to a new class.

Owing to the intrinsic limitations of γ-ray detection technique, however, the instruments' angular resolution has not yet reached the minimum level required to permit the transition from the unidentified limbo to the paradise of known objects, thus creating a continuing unidentified high-energy γ-ray source problem.

Indeed, the γ-ray sources first listed in the COS-B catalogue [1] are now hitting the "The Maximum Duration of Astronomical Incomprehension" estimated by Trimble [2] to be, on average, about 25 years.

2. The struggle towards identifications

To fully appreciate the peculiar situation of γ-ray astronomy, let us consider the parallel evolution of X and γ-ray astronomies, two space-based disciplines which started, more or less together, in the sixties. It is enlightening to compare the results of the first two seminal missions, belonging to the same series of NASA Small Astronomical Satellites and both launched into equatorial orbit from the Italian Malindi station, in Kenya.

SAS-1, better known as Uhuru, was active from 12 Dec 1970 to March 1973. Equipped with 840 cm^2 proportional counters, sensitive to photons with energies 2-20 keV, provided the first comprehensive and uniform all sky survey down to a sensitivity of 10^{-3} Crab.

SAS-2 was active from 19 November 1972 to 8 June 1973, when the transmitter ceased to function. Equipped with a 540 cm^2 spark chamber, sensitive to photons with energies from 20 MeV to 1 GeV, SAS-2 provided a partial sky survey at a sensitivity level corresponding to about half the Crab flux.

Since the Crab happens to be one of the brightest sources both in the X and in the γ-ray sky, the SAS-1/SAS-2 comparison illustrates from the start the difficulties of γ-ray astronomy, hampered by small number of photons, high background, low cross-section for detection, no possibility for collimation, to list just a few.

Uhuru produced a catalogue of 339 sources [3] while SAS-2 had a meagre harvest of three, plus the diffuse emission from the galactic plane and the extragalactic radiation [4]. X-ray sources were identified with different classes of objects – SNRs, binaries with compact companions, AGNs, clusters – while two of the three γ-ray sources were identified as pulsars.

Up to 1975 (when SAS-3 and COS-B went into orbit for a new generation of X-and γ-ray astronomy, respectively), 206 of the 339 Uhuru sources were without identification, while γ-ray astronomy had just 1 unidentified source, known as γ195+5, later to become Geminga.

In fact, would take 20 more years for γ-ray astronomy to build a catalogue similar to the Uhuru one, considering both the number of entries and the percentage of identification.

Meanwhile, X-ray astronomy had virtually dealt with the problem of the unidentified sources: with few notable exceptions, all X-ray sources (today orders of magnitude more numerous than in the Uhuru time) do now have a counterpart. This was achieved through a dramatic improvement of the instrument angular resolution, made possible by the development of the X-ray focussing technique.

First flown in 1978, on the second of NASA's three High Energy Astrophysical Observatories, HEAO-2, renamed Einstein after the launch, X-ray mirrors have been used in all subsequent missions and the Chandra performances can now rival with optical arc-sec standards. Moreover, extensive efforts on source identification have provided X-ray astronomers with additional tools, such as the F_x / F_{opt} parameter [5], which was found valuable in characterizing source families.

Gamma-ray astronomy has had a less glamorous development: more instruments were flown but the detection technique remained the same, thus providing similar performances in term of angular resolution. The number of sources grew in proportion to the number of photons detected and went from 25 at the end of the COS-B mission [1,6] to 272 and of the end of the EGRET mission [7]. With error boxes covering about 1 sq degree, unidentified sources accounts for 2/3 of the total EGRET harvest, roughly the same percentage as Uhuru. With such a poor angular resolution, γ-ray source identification must rely on additional pieces of information.

2.1. *Time signatures*

Since γ-ray astronomy is based on single photon detection, given a good enough clock chain (orbit-to-ground), accurate timing is always possible. Even with poorly determined positions, detecting in γ-rays variabilities known at other wavelengths is a powerful (and robust) identification tool.

Exploiting the timing of pulsars yielded the first γ-ray source identification back in 1975 with the detection of the Crab and Vela pulsar [4]. Although promising, the method has proven to be of somewhat limited use. As of today, only six out of 80 low latitude γ-ray sources have been confidently identified as pulsars [8]. More pulsars are certainly contributing to the galactic γ-ray source population: possibly they are already present as low statistical significance detections, alternatively they are still waiting to be discovered. Indeed, a few promising radio pulsars have been discovered inside EGRET source error boxes after the end of the mission. However, uncertainties in their timing history has made it impossible to search for their time signature in the γ-ray data. Globally, it is true to say that the vast majority of low galactic latitude sources is still unidentified.

Variability studies can be applied to all class of sources, provided they do show some kind of variability at any wavelength. Detection of simultaneous variability in γ-ray as well in the radio, optical or X-ray wavelengths has been a powerful tool to identify dozens of AGNs. Indeed, this class is so well established that radio-loud AGNs inside EGRET error boxes are considered a priori serious candidate counterparts, even if they do not exhibit significant variations.

No matter how powerful, however, variability alone cannot provide identification for all sources, a good fraction of which is remarkably constant.

2.2. *Multiwavelength approach*

In order to bridge the resolution gap, multiwavelength coverage of γ-ray error boxes has been extensively performed for the last 25 years. Although quite a number of searches were carried out in radio (spurred by the pulsar findings) the wavelength of election has been the X-ray one, both owing to the availability of focusing instrument and to the physical proximity of the γ-X energy intervals.

The paradigm of the current multiwavelength strategy is provided by the chase for Geminga. The sequence of observations, going from γ-rays to X-rays, to the optical and then back to γ -rays, shows that the interplay between the findings at different wavelengths is of paramount importance (see 9 for a review of the Geminga chase). After the discovery of the periodicity in X and γ–rays, the final piece of evidence came from the improvement in the γ-ray light curve generated by the use of the optical position and proper motion data [10].

However gratifying, Geminga's identification is the only success story we can tell so far. A number of potential (sometime compelling) candidates have been discovered by radio and X-ray telescopes (see e.g. 11 for a review of the proposed identifications) but, without an operating γ-ray telescope, the clinching evidence, yet again a time signature, is still lacking.

LSI 61°303 may be the next object to become identified as a γ-ray source: it has been on the waiting list since the beginning of the COS-B mission. The Einstein Observatory search for an X-ray counterpart of the gamma ray source 2CG 135+1 yielded the discovery of the X-ray emission from the variable radio star GT 0236 [12]. Deep scrutiny in γ-ray brought no definitive results, neither with COS-B nor with EGRET [13]. The flux of the source is too faint and the observations too scanty to allow for the detection of the 26 day radio periodicity, the smoking gun of this association.

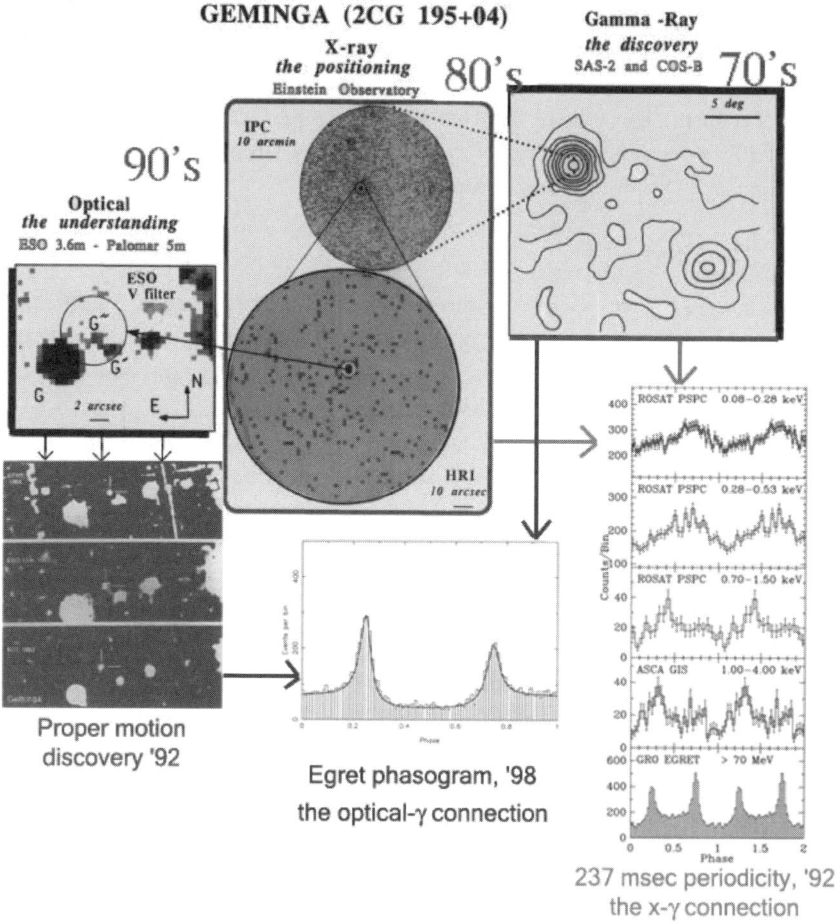

Figure 1. Summary of the multiwavelength results obtained over more than 20 years of Geminga observations. X-ray imaging, performed by the Einstein observatory, unveiled the potential X-ray counterpart for which an optical identification was also proposed. In 1992 the ROSAT X-ray satellite found the 237 msec source periodicity which was seen to be present also in gamma rays, thus linking the X-ray source to the γ-ray one (green arrows). Later, using the precise optical position, as well as the optical proper motion, the γ-ray light curve was seen to significantly improve, thus linking the optical source with the γ-ray one (blue arrows).

The chase for Geminga provided the first evidence for the existence for a new class of isolated neutron stars, the radio quiet ones [14].

The much awaited clue is now coming from the Cerenkov telescopes, heralding, maybe, a new era with space-borne and ground-based γ-telescopes working side by side to provide source identification. The Magic experiment (see Turini, these proceedings) has clearly detected the source periodic variability, making LSI 61°303 a very high energy γ-ray source, thus strengthening the 25 year old proposal [12] for the source identification.

3. The future

GLAST [see McEnery, these proceedings] promises a significant step forward in γ-ray astronomy, owing both to its much improved sensitivity and angular resolution. Thousands of sources positioned within few arcmins are expected from the first year all-sky survey.

Clearly, a Geminga-like multiwavelength approach cannot be applied to each source. A different strategy is being worked out by the GLAST Working Group devoted to the study of the Unidentified Sources, by proposing the idea of going "from detection to association to identification", mostly (but not only) through a statistical approach .

Even if the GLAST error boxes will be significantly smaller than the EGRET ones, multiple positional coincidences with field objects will be the rule rather than the exception. Thus, a "figure of merit" approach has to be developed for ranking the proposed candidates. First, for each source class a chance occurrence probability parameter should be computed in order to weight the relative abundance of a given source class. In general, counterparts belonging to a class of "certified" γ-ray emitters will be ranked higher than a counterparts belonging to a new, yet unseen, class. Moreover, the displacement of the proposed candidate from the best source position will come into play. Next, the energetic plausibility of the association should be evaluated. A proposed counterpart unable to meet the energetic requirement set by its γ-ray flux and its supposed distance will be discarded. Finally, the general source phenomenology will be scrutinized to see if its parameters, such as γ-ray emission efficiency, spectral shape, variability etc, are consistent with those of a given source class.

The time dimension will continue to be a very important asset for source identification. The use of time variabilities as an identification tool will be exploited also through a comprehensive program of coordinated observations devoted to different classes of objects.

Multiwavelength campaigns will be the last resort used, when everything else has failed or there is no other way to confirm an identification.

In order to avoid an a priori limitation on the GLAST potential for discovery, one should keep in mind that any figure of merit approach as well as any kind of monitoring campaign will be biased toward known source classes. Special care should be devoted to save the diversity (and unpredictability) of the γ-ray sky and of its blossoming species.

References

1. Swanenburg, B.N., et al Ap. J. 243,L69 (1981)
2. Trimble, V., AIPC 836, 3 (2006)
3. Forman, W., et al. Ap.J.S, 38, 357 (1978)
4. Fichtel, C.E. et al. Ap.J. 198, 163 (1975)
5. Zickgraf, F.-J. et al. A&A 406, 535 (2003)
6. Bignami, G.F., & Hermsen, W., Ann. Rev. A.A. 21, 79 (1983)
7. Hartman, R.C. et al Ap.J.S. 123,79 (1999)
8. Thompson, D. AIP Proceedings, volume 558. Edited by Felix A. Aharonian and Heinz J. Völk., ISBN 1-56396-990-4, p.103 (2001)
9. Bignami, G.F., & Caraveo, P.A., Ann. Rev. A.A. 34, 331 (1996)
10. Mattox, J.R., Halpern, J.P., Caraveo, P.A, Ap.J. 493,891 (1998)
11. La Palombara, N., et al. 458, 245 (2006)
12. Bignami, G.F., et al Ap.J. 247,P.L85 (1981)
13. Tavani, M., et al Ap.J. 497,L89 (1998)
14. Caraveo, P.A., Bignami, G.F., Truemper, J., A & A Review, 7, 209 (1996)

SUPERNOVAE AND GAMMA-RAY BURST

M. DELLA VALLE*

INAF-Arcetri Astrophysical Observatory
Largo E. Fermi, 5, Firenze, Italy
** E-mail: massimo@arcetri.astro.it*

We review the observational status of the Supernova/Gamma-Ray Burst connection. Present data suggest that a significant fraction of long-duration GRBs (although not all of them) are associated with bright SNe of type Ib/c. Current estimates of SN and GRB rates yield ratios GRB/SNe-Ibc in the range $\sim 0.4\% - 3\%$. In the few SN/GRB associations so far discovered the SN and GRB events appear to be simultaneous events. For the association GRB 060218/SN 2006aj, X and optical observations suggest that the SN and GRB are coeval events within ~ 0.1 days.

1. Introduction

A decade of observations of gamma-ray bursts (GRBs) has allowed to link a significant fraction of long-duration GRBs with the death of massive stars. This result is based on three pieces of evidence: i) there are four cases of association between "broad lined" supernovae (i.e. SNe-Ib/c characterized by a large kinetic energy, often labeled as hypernovae, HNe hereafter) and GRBs: GRB 980425/SN 1998bw (Galama et al. 1998), GRB 030329/SN 2003dh (Stanek et al. 2003, Hjorth et al. 2003), GRB 031203/SN 2003lw (Malesani et al. 2004) and GRB 060218/SN 2006aj (Campana et al. 2006; Pian et al. 2006); ii) in a few cases, spectroscopic observations of bumps observed during the late decline of GRB afterglows (Bloom et al. 1999) have revealed the presence of SN features in GRB 021211/SN 2002lt (Della Valle et al. 2003) and GRB 050525A/SN 2005nc (Della Valle et al. 2006a) and in XRF 020903 (Soderberg et al. 2005); iii) long GRBs are located inside star forming galaxies (Djorgovski et al. 1998, Fruchter et al. 2006). The standard theoretical scenario (e.g. Woosley & Bloom 2006) suggests that long duration GRBs are produced in the collapse of the core of H/He stripped-off massive stars, with an initial mass higher than $\sim 20\ M_\odot$, resulting in bright SNe-Ibc. However this scenario has been recently challenged by observations of GRB 060614 (Della Valle et al. 2006b, Fynbo et al. 2006, Gal-Yam et al. 2006), which show the existence of a sub-class of long-duration GRBs without an accompanying (bright) SN.

2. GRB 980425 and SN 1998bw

SN 1998bw was the first SN discovered spatially and temporally coincident with a GRB (Galama et al. 1998). GRB 980425 was discovered not at cosmological distances, as expected, but in the nearby galaxy ESO 184-G82 at $z = 0.0085$. This fact implied that GRB 980425 was underenegetic by 4 orders of magnitudes with respect to typical "cosmological" γ–budget of about 10^{51} erg. Also the evolution of SN 1998bw was peculiar (Patat et al. 2001). It was very bright at maximum ($M_V \sim -19$), the ejecta exhibited unusual high expansion velocities (about 30,000 km/s) and the radio emitting region associated with the GRB-SN was expanding at mildly relativistic velocities ($\Gamma \sim 2$, Kulkarni et al. 1998).

The theoretical modeling of the light curve and spectra suggests that SN 1998bw can be well reproduced by an extremely energetic explosion of an envelope-stripped star, which originally was $\sim 40 M_\odot$ on the main sequence (e.g. Woosley et al. 1999, Nakamura et al. 2001). This picture is consistent with the radio properties of SN 1998bw, which can be explained as due to the interaction of a mildly relativistic shock with a dense circumstellar medium (Tan et al. 2001, Weiler et al. 2002) due to the massive progenitor wind. Maeda et al. (2006) have recently modeled the bolometric lightcurve of SN 1998bw by assuming an high degree of asphericity in the SN explosions (see also Höflich, Wheeler & Wang 1999) which decreases the demand for an overproduction of ^{56}Ni ($\sim 0.4\ M_\odot$ ^{56}Ni) and would allow SN 1998bw to have an explosion energy of $\sim 10^{52}$ erg which is still larger than exhibited by "standard" SNe-Ibc, but not unusual for SNe associated with GRBs.

3. GRB 030329/SN 2003dh and GRB 031203/SN 2003lw

The association between two peculiar astrophysical objects such as GRB 980425 and SN 1998bw, was not taken by the GRB community as a proof for the existence of a general SN/GRB connection. The breakthrough in the study of the GRB/SN association arrived with the discovery of two nearby GRBs, namely GRB 030329 and GRB 031203.

GRB 030329 was discovered by the HETE–2 satellite at a redshift

$z = 0.1685$ (Greiner et al. 2003). SN features were detected in the spectra of the afterglow by several groups (Stanek et al. 2003, Hjorth et al. 2003; Kawabata et al. 2003; Matheson et al. 2003) and the associated SN (2003dh) looked strikingly similar to SN 1998bw. The gamma-ray and afterglow properties of this GRB were not unusual among GRBs, and therefore, the link between GRBs and SNe was eventually established to be more general. The modeling of the early spectra of SN 2003dh (Mazzali et al. 2003) has shown that SN 2003dh had a high explosion kinetic energy, $\sim 4 \times 10^{52}$ erg (if spherical symmetry is assumed). However, the light curve derived from fitting the spectra suggests that SN 2003dh was not as bright as SN 1998bw, ejecting only $\sim$ 0.3 $M_\odot$ of ^{56}Ni. The progenitor was a massive envelope-stripped star of $\sim 35 - 40 M_\odot$ on the main sequence. The spectral analysis of the nebular-phase emission lines carried out by Kosugi et al. (2004) suggests that the explosion of the progenitor of the GRB 030329 was aspherical, and that the axis is well aligned with both the GRB relativistic jet and our line of sight.

GRB 031203 was a 30s burst detected by the INTEGRAL alert system (Mereghetti et al. 2003) at $z = 0.1055$ (Prochaska et al. 2004). The burst energy was extremely low, of the order of 10^{49} erg, well below the "standard" $\sim 10^{51}$ erg of normal GRBs (Frail et al. 2001, Panaitescu & Kumar 2001) and similar to GRB 980425. However in this case, a very faint NIR afterglow could be discovered, orders of magnitude dimmer than usual GRB afterglows (Malesani et al. 2004). A few days after the GRB, a rebrightening was apparent in all optical bands (Thomsen et al. 2004; Cobb et al. 2004; Gal-Yam et al. 2004). The rebrightening amounted to $\sim$ 30% of the total flux (which is dominated by the host galaxy). Spectra of the rebrightening obtained on 2003 Dec 20 and Dec 30 (14 and 23 rest-frame days after the GRB) are remarkably similar to those of SN 1998bw obtained at comparable epochs. The light curve of SN 2003lw is also similar to that SN 1998bw, though broader by $\sim$ 10%. With the assumed reddening, SN 2003lw appears to be brighter than SN 1998bw by 0.3 mag in the V, R, and I bands. The absolute magnitudes of SN 2003lw are hence $M_V = -19.3 \pm 0.15$, $M_R = -19.5 \pm 0.1$, and $M_I = -19.5 \pm 0.1$. An analysis of early and late spectra (Mazzali et al. 2006) indicates that this Hypernova produced a large amount of Ni, possibly $\sim 0.5 - 0.6 M_\odot$. The progenitor mass could be as large as 40-50 $M_\odot$ on the main sequence.

4. GRB 060218 and SN 2006aj

GRB 060218 was detected with the burst alert telescope on-board Swift (Gehrels et al. 2006) at z=0.033. This burst was unusually long, with $T_{90} \sim 2100s$. The UVOT telescope found an emission peaking in a broad plateau first at UV wavelengths and later in the optical (see Fig. 2 in Campana et al. 2006). The lightcurve showed a minimum at about 200 ks after the gamma event and a rebrightening peaking at about 700 ks (due to the Ni-Co-Fe decay). A few days after the Swift observations, low resolution spectra (Pian et al. 2006) pointed out the presence of a rising SN (2006aj) with broad emission lines similarly to those observed in other GRB-SNe. The high energy spectrum soften with time and can be fitted with a power-law, as observed in other GRBs. The most striking feature exhibited by this gamma event (see Campana et al. 2006) is a thermal component present in the XRT data, up to 10 ks, and in the UVOT data, up to about 100 ks. This black body component shows a decreasing temperature accompanied by an increasing luminosity, which implies an increase in the apparent emission radius from an initial 5×10^{11} cm to about 3×10^{14} cm, in about 100ks. This corresponds to an expansion velocity of the order of 30,000 km/s, which is quite typical for GRB-SNe (see Patat et al. 2001). After assuming linear expansion one can estimate the star radius of the progenitors to be of the order of a few $\times 10^{11}$ cm. This is about one order of magnitude smaller than a blue supergiant star and comparable to the size of a Wolf-Rayet star.

5. Rates of SNe Ib/c, Hypernovae and GRBs

The fraction of SNe-Ibc that produces GRBs can be measured as follows. A rate of $\sim 2 \times 10^4$ SNe-Ibc Gpc^{-3} yr^{-1} is derived by combining the local density of B luminosity of $\sim 1.2 \times 10^8 L_{B,\odot}$ Mpc^{-3} (e.g. Madau, Della Valle & Panagia 1998) with the rate of 0.16 SNe-Ibc per century and per 10^{10} $L_{B,\odot}$ (SNu units, Cappellaro, Evans & Turatto 1999) measured in Sbc–Irr galaxies. This SN rate has to be compared with the rate of "cosmological" GRBs of ~ 1 GRB Gpc^{-3} yr^{-1} (Guetta, Piran & Waxman 2005, Schmidt 2001) after rescaling for the jet beaming factor, f_b^{-1}. There exist different estimates for this parameter: from ~ 75 (Guetta, Piran & Waxman 2005) to ~ 500 (Frail et al. 2001) corresponding to beaming angles $\sim 10°$–$4°$, respectively. Taking these figures at their face value, we find this ratio to be in the range: $\sim 0.4\% - 3\%$. Recently Amati et al. (2006) have convincingly shown that GRB 060218 can be the prototype of a class of truly sub-energetic GRBs. Several authors (e.g., Guetta et al. 2004, Della Valle 2006,

Pian et al. 2006, Soderberg et al. 2006a, Cobb et al. 2006) have pointed out that, since the volume sampled at $z = 0.033$ is $\sim 10^6$ times smaller than that probed by classical, distant GRBs, the occurrence of two events such as GRB 980425 and GRB 060218 within $\sim$ 150 Mpc distance may implies that sub–energetic GRBs may be the most frequent gamma-ray events in the Universe. After assuming that truly sub-energetic GRBs are much less collimated events than "cosmological" GRBs ($f_b^{-1} \sim 1-2$ for GRB 060218, Soderberg et al. 2006a), Guetta & Della Valle (2006) have derived a ratio "faint GRBs/SNe-Ibc: $\sim 1\% - 9\%$, which may be larger than the ratio "cosmological" GRBs / SNe-Ibc reported above ($0.4\% - 3\%$), but yet not so dramatically different to call for two different GRB populations. Radio surveys give independent and consistent constraints: Berger et al. (2003) find that the incidence of SN 1998bw-like events, in the nearby universe, is $< 3\%$, while Soderberg et al. (2006b) find "cosmological" GRB/SNe-Ibc $< 10\%$.

6. Conclusions

GRB observations of the past eight years have produced an astonishing advance in the understanding of the GRB-SN phenomenon. Although many questions are still open (e.g. Della Valle 2006, Woosley and Bloom 2006) a number of robust conclusions may be drawn:

i) A significant fraction of long duration GRBs originates from the death of massive stars. This fact is well documented by: i) the direct observations of four SNe associated with GRBs; ii) the theoretical modeling of their spectra and lightcurve which suggest progenitor stars more massive (on the main sequence) than $\sim 20M_\odot$.

ii) Observations of GRB 060218 coupled with simple theoretical arguments (see Campana et al. 2006) indicate that the progenitor star of the associated SN (2006aj) had a radius of about $\sim 5 \times 10^{11}$ cm. This is similar to the size of a Wolf-Rayet star and fully consistent with the fact that all GRB-SNe, so far observed, belong to Ib/c types, i.e., they derive from the collapse of H/He stripped-off massive stars.

iii) The near simultaneous observations of the non thermal (GRB) and thermal (SN) X-ray emissions in the GRB 060218/SN 2006aj association, indicate that the SN and the GRB are coeval events within $\sim$ 0.1 day. This is a result with important theoretical implications. Before the occurrence of GRB 060218 the simultaneity between GRB and SN events was verified within a few days (see Della Valle 2006).

iv) Only $0.4\% - 3\%$ of SNe-Ibc (corresponding to less than 1% of all core-collapse SNe) are capable to produce GRBs. Therefore some special circumstances are requested to allow a massive star to become a GRB progenitor. Recent theoretical studies indicate that rotation (Woosley & Hegel 2006; Yoon & Langer 2005), metallicity (Fruchter et al. 2006), binarity (Podsiadlowski et al. 2004, Mirabel et al. 2003) may play an important role.

v) Recent observations of GRB 060614 (Della Valle et al. 2006b, Fynbo et al. 2006, Gal-Yam et al. 2006) challenge the neat idea that all long-duration GRBs are produced in bright Supernova explosions. Indeed any SN associated with this GRB was at least 100 times fainter (in R band) than the other SNe associated with GRBs. This fact suggests that GRB 060614 is the proto-type of a new class of long-duration GRB which are produced in a new kind of massive star death, different from those producing bright SNe-Ibc.

References

Amati, L., Della Valle, M., Frontera, F., Malesani, D., Guidorzi, C., Montanari, E., Pian, E. 2006, A&A, in press
Berger, E., Kulkarni, S. R., Frail, D. A., Soderberg, A. M. 2003, ApJ, 599, 408
Bloom, J. S., Kulkarni, S. R., Djorgovski, S. G. et al. 1999, Nature, 401, 453
Campana, S. et al. 2006, Nature, 442, 1008
Cappellaro, E., Evans, R., Turatto, M. 1999, A&A, 351, 459
Cobb, B. E., Baylin, C. D., van Dokkum, P. G., Buxton, M. M. & Bloom, J. S. 2004, ApJ, 608, L93
Cobb B. E., Bailyn, C. D., van Dokkum, P. G., Natarajan, P. 2006, ApJ, 645, L113
Della Valle M., Malesani, D., Benetti, S. et al. 2003, A&A, 406, L33
Della Valle, M. 2006, Proceedings of the 16th Annual October Astrophysics Conference in Maryland, "Gamma Ray Bursts in the Swift Era", eds. S. Holt, N. Gehrels and J. Nousek (astro-ph/06004110)
Della Valle, M., Malesani, D., Bloom, J. et al. 2006, ApJ, 642, L103
Della Valle, M., Chincarini, G., Panagia, N., et al. 2006b, Nature, accepted (astro-ph/0608322)
Djorgovski, S. G., Kulkarni, S. R., Bloom, J. S., Goodrich, R., Frail, D. A., Piro, L.; Palazzi, E. 1998, ApJ, 508, L17
Frail, D. A., Kulkarni, S. R., Sari, R. et al. 2001, ApJ, 562, L55
Fruchter, A., et al. 2006, Nature, 441, 463
Fynbo, P.U., Watson, D., Thoene, C.C., et al. 2006, Nature, accepted (astro–ph/0608313)
Gal-Yam, A., Moon, D.S., Fox, D.B. et al. 2004, ApJ, 609, L59
Gal-Yam, A., Fox, D., Price, P., et al. 2006, Nature, accepted (astro–ph/0608257)
Galama, T.J., Vreeswijk, P.M., van Paradijs, J., et al. 1998, Nature, 395, 670
Gehrels, N., Chincarini, G., Giommi, P., et al. 2004, ApJ, 611, 1005

Greiner, J. et al. 2003, GCN 2020
Guetta, D., Perna, R., Stella, L., Vietri, M. 2004, ApJ, 615, L73
Guetta, D., Piran, T., Waxman, E. 2005, ApJ, 619, 412
Guetta, D. & Della Valle, M. 2006, ApJ, submitted
Hjorth J., Sollerman, J., Moller, P. et al. 2003, Nature,423, 847
Höflich, P.; Wheeler, J. C.; Wang, L. 1999, ApJ, 521, 179
Kawabata, K. S., Deng, J., Wang, L. et al. 2003, ApJ, 593, L19
Kosugi, G. Mizumoto, Y. Kawai, N. et al. 2004, PASJ, 56, 61
Kulkarni, S. R., Frail, D. A., Wieringa, M. H. et al. 1998, Nature, 395, 663
Madau, P., Della Valle, M., Panagia, N. 1998, MNRAS, 297, L17
Maeda, K., Mazzali, P., Nomoto, K. 2006, ApJ, 645, 1331
Malesani D. Tagliaferri, G., Chincarini, G. et al.2004, ApJ, 609 L5
Matheson, T.; Garnavich, P. M.; Stanek, K. Z. et al. 2003, ApJ, 599, 394
Mazzali, P., Deng, J., Tominaga, N. et al. 2003, ApJ, 599, L95
Mazzali et al. 2006, ApJ, 645, 1323
Mereghetti, S., & Götz, D. 2003, GCN Circ. 2460
Mirabel, I, F. 2004, RMxAC, 20, 14
Nakamura, T., Mazzali, P., Nomoto, K., Iwamoto, K. 2001, ApJ, 550, 991
Panaitescu, A., Kumar, P. 2001, ApJ, 560, L49
Patat, F., Cappellaro, E., Danziger, I.J., et al. 2001, ApJ, 555, 900
Pian, E., et al. 2006, Nature, 442, 1011
Podsiadlowski, P., Mazzali, P., Nomoto, K., Lazzati, D., Cappellaro, E. 2004, ApJ, 607, L17
Prochaska, J.X., Bloom, J. S., Chen, H., Hurley, K, C., Melbourne, J., Dressler, A., Graham, J.R., Osip, D. J., Vacca, W.D. 2004, ApJ, 611, 200 Schmidt, M. 2001, ApJ, 552, 36
Soderberg, A. M., et al. 2005, ApJ, 627, 877
Soderberg, A. M. et al. 2006a, Nature, 442, 1014
Soderberg, A.M., Nakar, E., Kulkarni, S.R., & Berger, E. 2006b, ApJ, 638, 930
Stanek, K. Z., Matheson, T., Garnavich, P. M. et al. 2003, ApJ, 591, L17
Tan, J.C., Matzner, C.D., McKee, C.F. 2001, ApJ, 551, 946
Thomsen, B., Hjorth, J., Watson, D. et al. 2004, A&A, 419, L21
Weiler, K.W., Panagia, N., Montes, M.J., Sramek, R.A. 2002, ARA&A, 40, 387
Woosley, S.E., Eastman, R.G., Schmidt, B.P. 1999, ApJ, 516, 788
Woosley, S.E. & Heger, A. 2006, ApJ, 637, 914
Woosley, S.E. & Bloom, J. 2006, ARA&A, 44, 507
Yoon, S.C. & Langer, N. 2005, A&A, 443, 643

FIRST CYCLE OF MAGIC GALACTIC OBSERVATIONS

J. CORTINA for the MAGIC COLLABORATION

Institut de Fisica d'Altes Energies,
Edifici Cn, Campus UAB,
Bellaterra, 08193 Barcelona, Spain
E-mail: cortina@ifae.es

I review the results of the galactic observations performed during the first 15-month operation cycle of MAGIC Very High Energy (VHE) Gamma Ray telescope. Highlights of this first cycle are the study of the Crab Nebula and its pulsar, and two unidentified HESS sources (in particular the tentative identification of HESS J1834-087 with a molecular cloud), the measurement of the Galactic Center energy spectrum up to 10 TeV and the discovery of the variable gamma-ray binary LS I +61 303.

1. Introduction: The MAGIC Telescope

MAGIC is a telescope for VHE ($E \geq 10$ GeV) γ-ray observation (for more details see [1] in these proceedings). MAGIC's sensitivity above 100 GeV is $\sim 2.5\%$ of the Crab nebula flux in 50 hours of observations. The energy resolution above 200 GeV is better than 30%. The angular resolution is $\sim 0.1^\circ$, while source localization in the sky is provided with a precision of $\sim 2'$.

The physics program developed with the MAGIC telescope includes both, topics of fundamental physics and astrophysics. In this paper we present the results regarding the observations of galactic targets. The results from extragalactic observations are presented elsewhere in these proceedings [2,3].

2. Highlights of cycle I

MAGIC's first observation cycle spanned the period from January 2005 to April 2006. About one fourth of the observation time (excluding that devoted to Crab nebula technical observations) was devoted to galactic objects. The observations covered both candidates and well established VHE γ-ray emitters, and included the following types of objects: supernova remnants (SNRs), pulsars, pulsar wind nebulae (PWN), microquasars (μQSRs), the Galactic Center (GC), one unidentified TeV source and one cataclysmic variable. In this section we highlight the results obtained so far from such observations, and concentrate, in section 3, on the most interesting case of the γ-ray binary LS I +61 303.

2.1. *The Crab nebula and pulsar*

The Crab nebula is a steady emitter at GeV and TeV energies, what makes it into an excellent calibration candle. This object has been observed extensively in the past over a wide range of wavelengths, up to nearly 100 TeV. Nevertheless, some of the relevant physics phenomena are expected to happen in the VHE domain, namely the spectrum showing an inverse Compton (IC) peak close to 100 GeV, a cut-off of the pulsed emission somewhere between 10 and 100 GeV, and the verification of the flux stability down to the percent level. The existing VHE γ-ray experimental data is well described by electron acceleration followed by the IC scattering of photons generated by synchrotron radiation (synchrotron self Compton process).

Along the first cycle of MAGIC's regular observations, a significant amount of time has been devoted to observe the Crab nebula, both for technical and astrophysical studies. A sample of 12 hours of selected data has been used to measure with high precision the spectrum down to $\sim$100 GeV, as shown in figure 1 [4]. We have also carried out a search for pulsed γ-ray emission from Crab pulsar, albeit with negative results. The derived upper limits (95% C.L.) are 2.0×10^{-10} ph s^{-1}cm^{-2} at 90 GeV and 1.1×10^{-10} ph s^{-1}cm^{-2} at 150 GeV.

2.2. *Supernova Remnants*

Shocks produced at supernova explosions are assumed to be the source of the galactic component of the cosmic ray flux [5]. The proof that this is the case could be provided by observations in the VHE domain. The rationale is that the hadronic component of the cosmic rays –enhanced close to their source, i.e. the SNR– should produce VHE γ-rays by the interaction with nearby dense molecular clouds. Although recent data seem to indicate that this is the case, it is difficult to disentangle the VHE component initiated by hadrons from that produced by Bremsstrahlung and IC processes

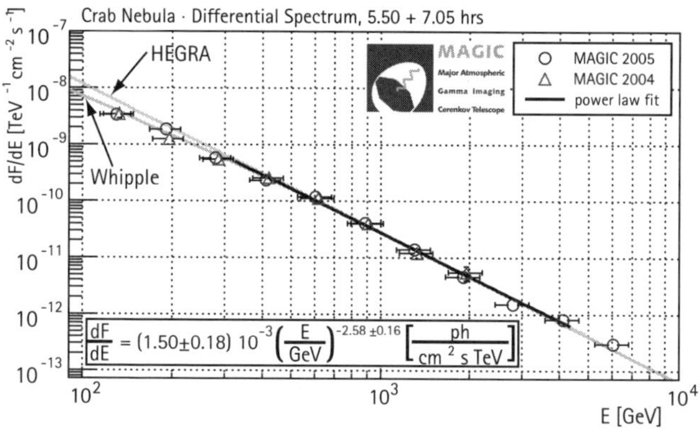

Fig. 1. Energy spectrum above 100 GeV from the Crab nebula measured by MAGIC in two different observation seasons. For comparison, the extrapolations down to 100 of HEGRA and Whipple measurements are shown.

by accelerated electrons. Therefore more data in the TeV regime together with multi-wavelength studies are needed to finally solve the long-standing puzzle of the origin of galactic cosmic rays.

Within its program of observation of galactic sources, MAGIC has observed a number of supernova remnants. In particular, we are observing several of the brightest EGRET sources associated to SNRs, and the analysis of the acquired data is in progress. On the other hand, we have confirmed the VHE γ-ray emission from the SNRs HESS J1813-178 [6] and HESS J1834-087 (W41) [7]. Our results have confirmed SNRs as a well established population of VHE γ-ray emitters. The energy spectra measured by MAGIC are both well described by an unbroken power law and an intensity of about 10% of the Crab nebula flux. Furthermore, MAGIC has proven its capability to study moderately extended sources by observing HESS J1834-087. Interestingly, the maximum of the VHE emission for this object has been correlated with a maximum in the density of a nearby molecular cloud. Although the mechanism responsible for the VHE radiation remains yet to be clarified, this is a hint that it could be produced by high energy hadrons interacting with the molecular cloud.

2.3. *Galactic Center*

We have also measured the VHE γ-ray flux from the GC [10]. The possibility to indirectly detect dark matter through its annihilation into VHE γ-rays has risen the interest to observe this region during the last years.

Our observations have confirmed a point-like γ-ray excess whose location is spatially consistent with Sgr A* as well as Sgr A East. The energy spectrum of the detected emission is well described by an unbroken power law of photon index $\alpha = -2.2$ and intensity about 10% of that of the Crab nebula flux at 1 TeV. The power law spectrum disfavours dark matter annihilation as the main origin of the detected flux. There is no evidence for variability of the flux on hour/day time scales nor on a year scale. This suggests that the acceleration takes place in a steady object such as a SNR or a PWN, and not in the central black hole.

3. The γ-ray binary LS I +61 303

3.1. *Description of the object*

One of the most studied μQSR candidates is LS I +61 303. This system is composed of a compact object of unknown nature (neutron star or black hole) in a highly eccentric ($e = 0.7$) orbit around a Be star. The orbital period –with associated radio [12] and X-ray [13] outbursts– is 26.5 days and periastron passage is at phase 0.23 [14].

LS I +61 303 is also one of the two μQSR candidates positionally coincident with EGRET γ-ray sources [16], and the only one located in the Northern Hemisphere –hence a suitable target for MAGIC. However, the large uncertainty of the position of the EGRET source has not allowed an unambiguous association with LS I +61 303.

3.2. *MAGIC observations*

LS I +61 303 was observed in the VHE regime with MAGIC during 54 hours (after standard quality selection, discarding bad weather data) between October 2005 and March 2006 [17]. The data analysis was carried out using the standard MAGIC reconstruction and analysis software [6,7,10]. Figure 2 shows the reconstructed γ-ray map during two different observation periods, around periastron passage and at higher (0.4-0.7) orbital phases. No significant excess in the number of γ-ray events is detected around periastron passage, whereas it shows up clearly (9.4σ statistical significance) at later orbital phases. The distribution of γ-ray excess is consistent with a point-like source at a position which is in agreement with the position of LS I +61 303. In the natural case in which the VHE emission is produced by the same object detected at EGRET energies, this result identifies a γ-ray source that resisted classification during the last three decades.

Our measurements show that the VHE γ-ray emission from LS I +61 303

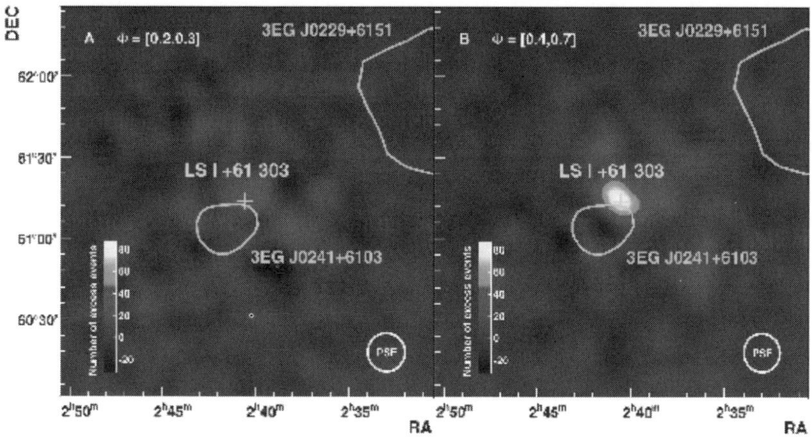

Fig. 2. Smoothed maps of γ-ray excess events above 400 GeV around LS I +61 303. (A) Observations over 15.5 hours corresponding to data around periastron (i.e., between orbital phases 0.2 and 0.3). (B) Observations over 10.7 hours at orbital phase between 0.4 and 0.7. The position of the optical source LSI +61 303 (yellow cross) and the 95% confidence level contours for the EGRET sources 3EG J0229+6151 and 3EG J0241+6103 (green contours) are also shown. The bottom right circle shows the size of the point spread function of MAGIC (1σ radius). From Albert *et al.* [17].

is variable. The γ-ray flux above 400 GeV coming from the direction of LS I +61 303 (see Figure 3) has a maximum corresponding to about 16% of the Crab nebula flux, and is detected at around phase 0.6. The combined statistical significance of the 3 highest flux measurements is 8.7σ, for an integrated observation time of 4.2 hours. The probability for the distribution of measured fluxes to be a statistical fluctuation of a constant flux (obtained from a χ^2 fit of a constant function to the entire data sample) is 3×10^{-5}. The fact that the detections occur at similar orbital phases hints at a periodic nature of the VHE γ-ray emission.

The VHE spectrum derived from data between ~200 GeV and ~4 TeV at orbital phases between 0.4 and 0.7 is fitted reasonably well by a power law function:

$$dN/(dA/dt/dE) = (2.7 \pm 0.4 \pm 0.8) \times 10^{-12} \times E^{(-2.6 \pm 0.2 \pm 0.2)} \quad \mathrm{cm^{-2}s^{-1}TeV^{-1}} \qquad (1)$$

where N is the number of γ-rays reaching Earth per unit area A, time t and energy E (expressed in TeV). Errors quoted are statistical and systematic, respectively. This spectrum is consistent with that measured by EGRET for a spectral break between 10 and 100 GeV. The flux from LS I +61 303

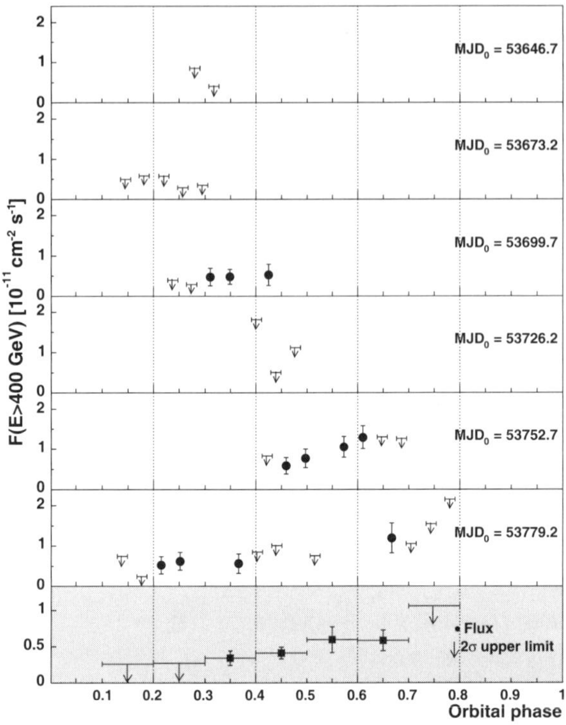

Fig. 3. VHE γ-ray flux of LS I +61 303 as a function of orbital phase for the six observed orbital cycles (six upper panels, one point per observation night) and averaged for the entire observation time (bottom panel). Vertical error bars include 1σ statistical error and 10% systematic uncertainty on day-to-day relative fluxes. The modified Julian date (MJD) corresponding to orbital phase 0 is indicated for every orbital cycle. From Albert *et al.* [17].

above 200 GeV corresponds to an isotropic luminosity of $\sim 7 \times 10^{33}$ erg s^{-1} (assuming a distance to the system of 2 kpc [18]), of the same order of that of the similar object LS 5039 [19], and a factor ~ 2 lower than the previous experimental upper limit ($< 8.8 \times 10^{-12}$ cm^{-2} s^{-1} above 500 GeV) [20]. LS I +61 303 displays more luminosity at GeV than at X-ray energies, a behavior shared also by LS 5039.

3.3. *Emission scenarios*

LS I +61 303 belongs, together with LS 5039 [19] and PSR B1259-63 [21], to a new class of objects, the so-called γ-ray binary systems, whose electromagnetic emission extends up to the TeV domain. LS I +61 303 and LS 5039 are usually thought to be μQSRs, since high-resolution radio imaging techniques have shown extended, radio-emitting structures with angular extension of $\sim$0.01 to $\sim$0.1 arc-sec, interpreted as a two-sided relativistic jet ($\beta/c = 0.6$) [15]. In such scenario, the high energy emission is produced by the shocks triggered at the relativistic jets [11]. However, no clear signal of the presence of an accretion disk (in particular an spectral feature between $\sim$ 10 and $\sim$ 100 keV due to the cut-off of the thermal emission) has been observed so far. Because of that, it has been alternatively proposed that relativistic particles could be injected into the surrounding medium by the wind from a young pulsar [22], which seems to be the case of PSR B1259-63. In the case of LS I +61 303 also the resemblance of the time variability and the radio/X-ray spectra with those of young pulsars seem to support such hypothesis. However, no pulsed emission has been detected from LS I +61 303. Therefore, the existing data in the radio/X-/γ-ray domain, cannot conclusively confirm, or rule out, none of the two proposed scenarios. Thus, the question of whether the three known γ-ray binaries produce TeV emission by the same mechanism, and by which one, is currently object of an intense debate [23].

The observation of jets has triggered the study of different microquasar-based γ-ray emission models, some regarding hadronic mechanisms: relativistic protons in the jet interact with non-relativistic stellar wind ions, producing γ-rays via neutral pion decay [24]; some regarding leptonic mechanisms: IC scattering of relativistic electrons in the jet on stellar and/or synchrotron photons [25].

The TeV flux measured by MAGIC has a maximum at phases 0.5-0.6 (see Figure 3), overlapping with the onset of the radio outburst. The timescale of the TeV variability constrains the emitting region to be smaller than a few 10^{15} cm (or 0.1 arc-sec at 2 kpc), which is larger than the binary system and of the same order of the detected radio jet-like structures. The maximum flux is not detected at periastron, when the accretion rate is expected to be the largest. VHE emission after periastron passage has been predicted considering electromagnetic cascading within the binary system [26] and due to the geometrical effect of changes in the relative orientation of the stellar companion with respect to the compact object and jet as it impacts the position and depth of the $\gamma\gamma$ absorption trough [27]. The existing data are, however, not conclusive to confirm or rule out any of the theoretical models existing in the literature.

LS I +61 303 is an excellent laboratory to study the VHE γ-ray emission and absorption processes taking place in massive X-ray binaries: the high eccentricity of the binary system provides very different physical conditions to be tested on timescales of less than one month. Future MAGIC observations will test both, the periodicity of the signal and its intra-night variability.

Acknowledgements. We thank the Instituto de Astrofísica de Canarias for the excellent working conditions at the Observatory Roque de los Muchachos in La Palma

References

1. Turini, N.: These proceedings.
2. Piccioli, A.: These proceedings.
3. Bigongiari, C.: These proceedings.
4. Wagner, R. *et al.*: Proceedings of the 29th International Cosmic Ray Conference, Pune (India) 3-11 Aug 2005, Vol. 4, 163-166.
5. Baade, W. and Zwicky, F.: Phys. Rev. **45** 138 (1934).
6. Albert, J. *et al.*: Astrophys. J. **637** L41-L44 (2006).
7. Albert, J. *et al.*, Astrophys. J. **643** L53-L56 (2006).
8. Dame, T. M. et al.,: Astrophys. J. **547** 792-813 (2001).
9. White, R. L., et al.: Astron. J. **130** 586-596. (2005).
10. Albert, J. *et al.*: Astrophys. J.: **638**, L101-L104 (2006).
11. Mirabel, I. F. and Rodrguez, L. F.: Annual Review of Astronomy and Astrophysics **37**, 409-443 (1999).
12. Gregory, P. C., Taylor, A. R.: Nature **272**, 704-706 (1978).
13. Taylor., A. R. *et al.*: Astron. Astrophys. **305**, 817-824 (1996).
14. Casares, J. *et al.*: Mon. Not. R. Astron. Soc. **360**, 1105-1109 (2005).
15. Massi, M. *et al.*: Astron. Astrophys. **414**, L1-L4 (2004).
16. Kniffen D. A. *et al.*: Astrophys. J. **486**, 126-131 (1997).
17. Albert, J. *et al.*: Science **312**, 1771-1773 (2006).
18. Frail, D. A. and Hjellming, R. M.: Astron. J. **101** 2126-2130 (1991).
19. Aharonian, F. et al., Science, **309**, Issue 5735, 746-749 (2005).
20. Fegan S. J. *et al.*: Astrophys. J. **624**, 638-655 (2005).
21. Aharonian, F. *et al.* Astron. Astrophys. **442** 1-10 (2005).
22. Maraschi, L. and Treves, A.: Mon. Not. R. Astron. Soc. **194**, 1P (1981).
23. Mirabel, F.: Science **312** 1759-1760 (2006).
24. Romero, G. E.: These proceedings.
25. Bosch-Ramon, V. et al.: Astron. Astrophys. **447**, 263-276 (2006).
26. Bednarek, W.: Mon. Not. R. Astron. Soc. **368**, 579-591 (2006).
27. Gupta, S. and Boettcher, M.: astro-ph/0606590.

Gamma-Rays and Neutrinos from a SNR in the Galactic Center

Vincenzo Cavasinni[1,2] †, Dario Grasso[2] ‡, Luca Maccione[3,4] *

[1] *Dip. di Fisica "E. Fermi", Università di Pisa, Largo B. Pontecorvo, 3, Pisa*
[2] *I.N.F.N. Sezione di Pisa*
[3] *S.I.S.S.A., Via Beirut, 2-4 I-34014, Trieste*
[4] *I.N.F.N. Sezione di Trieste*

† *Vincenzo.Cavasinni@pi.infn.it;* ‡ *dario.grasso@pi.infn.it;* * *maccione@sissa.it*

The recent detection of high energy gamma-ray emission from some galactic Supernova Remnants (SNRs) provided the first observational support to the long standing hypothesis that SNRs are the sources of galactic Cosmic Rays (CRs). The possible detection of neutrinos coming from the same sources would be valuable to confirm the hadronic origin of that radiation. Having this scenario in mind we evaluate the neutrino-photon flux ratio to be expected from pion decay in the TeV range. We applied our results to estimate the neutrino flux from the Galactic Centre region which was showed by HESS and MAGIC to be an active gamma-ray source, most likely powered by the SNR Sgr A East. According to our results that source should be detectable by a km^3 Mediterranean Neutrino Telescope.

1. Introduction

It is generally accepted that the bulk of Cosmic Rays (CRs) in our galaxy should be accelerated in the supersonic outflow of galactic SuperNova Remnants (SNRs). As a consequence, the CR energy density in the vicinity of a SNR should be significantly larger than in the rest of the galaxy. Furthermore, if an active SNR lies inside or close to a dense region (*embedded SNR*), e.g. a giant molecular cloud (GMC), the interaction between the relativistic particles and the dense gas ($n_H \sim 10^2 \div 10^6$ cm^{-3} in the case of GMCs) would lead to an enhanced production of γ-rays and neutrinos.[1,2] The existence of a correlation between the position of SNRs and that of giant molecular clouds (GMCs) is supported by both infrared observations and theoretical arguments. In fact, infrared observations confirmed the theoretical expectation that stars, especially massive short-living ones, form mainly in the womb of GMCs and can complete their evolution and explode within the parent GMC before this is swept-out by stellar winds.

Several satellites (in particular Egret[3]) have demonstrated the existence of a correlation between γ-ray emission and dense molecular clouds. However, those observations can be explained without invoking any local CR overdensity with respect to that observed at Earth. A significant progress has been achieved recently by the new generation of atmospheric Cherenkov telescopes. In particular, HESS was able to find the smoking-gun of the interaction of CRs accelerated by an active SNR with a close MC especially with the observations of RXJ1713.7-3946[4] and the very recent measurements of the Galactic Centre (GC) ridge emission.[5]

Although the arguments leading to the conclusion that those signals have hadronic origin are quite convincing, especially for the GC emission, the information provided by the detection of neutrinos from the same sources would be valuable to confirm our present understanding of this scenario.

Several Neutrino telescopes are already operating, or under construction, around the world.[6] Due to their large detection volume and observation time, these instruments may be able to detect at least few neutrinos from some SNRs, especially if they are close to a GMC. Since the energy threshold above which Neutrino Telescopes may identify astrophysical neutrino sources ($\gtrsim$ 100 GeV) is approximately the same above which atmospheric Cherenkov telescopes can detect photons, we might directly test if HE photons and neutrinos have the same hadronic origin confirming the long-standing hypothesis that HECRs are accelerated by SNRs. We will estimate here the expected neutrino/photon flux ratio in the TeV energy range from embedded SNRs, following the line already showed in [7,10].

2. Photon emissivity by π^0 decay.

We want to describe the process $pp \to NN + \mathrm{n}_\pi \pi^{\pm,0}$, where n_π is the pion multiplicity and N is either a proton or a neutron, and the subsequent pion decay.

The general expression of the photon emissivity is

$$Q_{\pi^0}(E_{\pi^0}) = c\, n_H \int_{E_p^{min}(E_{\pi^0})}^{E_p^{max}} dE_p\; n_p(E_p) \frac{d\sigma}{dE_{\pi^0}}(E_p, E_{\pi^0}) \;, \qquad (1)$$

where we assumed that hydrogen, with mean density n_H, is the main proton target. When the proton energy is much larger than the proton mass the differential pion production cross section can be approximated by a scaling expression,[13] as we verified[7] using the numerical package PYTHIA.[14]

In all cases in which the proton spectrum is well approximated by a power spectrum with spectral index α[a] the photon emissivity can be rewritten as

$$Q_\gamma(E_\gamma) \simeq \frac{2}{\alpha}\sigma_0 \, c \, n_H Y(\alpha) \, n_p(E_\gamma) \, , \tag{2}$$

where $Y(\alpha)$ is the pion yield as in [8].

3. Muon and Electron Neutrino emissivity.

Inelastic pp collisions lead to roughly the same number of π^0's, π^+'s and π^-'s . If π^0 decay is the dominant source of high energy photons, an emission of neutrinos has to be expected as well. The emissivity of muon neutrinos and antineutrinos produced by direct π^+ and π^-decay is

$$Q^{\pi}_{\nu_\mu}(E_{\nu_\mu}) \simeq \frac{m_\pi^2}{m_\pi^2 - m_\mu^2} \int_{E_\pi^{min}(E_\nu)}^{\infty} \frac{dE_\pi}{E_\pi} Q_{\pi^\pm}(E_\pi) \, , \tag{3}$$

where $Q_{\pi^\pm}(E_\pi) = Q_{\pi^0}(E_\pi) \times (1 \pm \epsilon)$. The ϵ represents the relative asymmetry between the positive and negative pions effectively produced in pp collisions:

$$\epsilon \equiv \frac{N_{\pi^+} - N_{\pi^-}}{N_{\pi^+} + N_{\pi^-}} \, . \tag{4}$$

The parameter ϵ has been estimated in [7] by a simulation performed with the help of PYTHIA. The result is shown in Fig. 1. We can now write

$$Q^{\pi}_{\nu_\mu}(E_\nu) \simeq \left(\frac{1}{1 - r^2}\right)^{1-\alpha} \frac{1 \pm \langle\epsilon\rangle}{2} Q_\gamma(E_\nu) \, , \tag{5}$$

where $\langle\epsilon\rangle$ represents the average of ϵ over the relevant energy range $E \in [10 \text{ GeV} - 500 \text{ TeV}]$.

The computation of the contribution of secondary muon decay $\mu \to e\nu_\mu\nu_e$ to the neutrino emissivity is slightly more involved. The detailed calculation of the dynamics has been done in [7] following [15]. The kaon contribution to the neutrino emissivity comes out to be few percent of the pion one.[7] Finally, using a mean asymmetry ϵ of about 5%, we find the values of the ratios Q_{ν_i}/Q_γ which we show in Tab. 1 for several values of the spectral index α.

---

[a]Notice that by writing the proton spectrum as a pure power law, we implicitly assume that the ultra-violet cut-off is well above the maximal energy relevant for γ-ray observations. In the case of the GC this is well motivated by the energy spectra observed by HESS which is steady power laws at least up to ~ 6 TeV.[17,19]

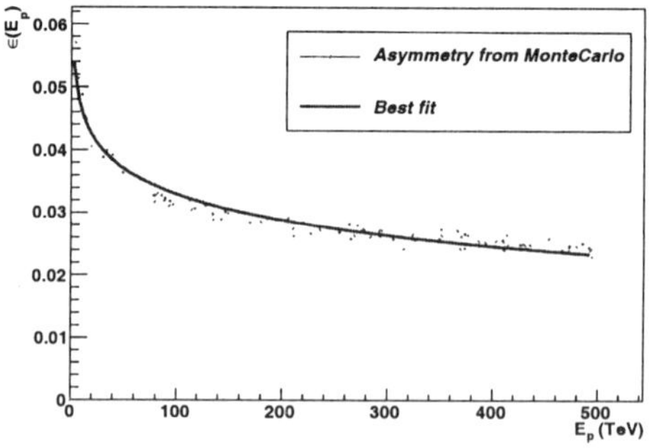

Fig. 1. Energy dependence and best fit of ϵ, as provided by PYTHIA, taken from [7].

Table 1. ν/γ emissivity ratios varying α.

Q_{ν_i}/Q_γ	$\alpha = 2.0$	$\alpha = 2.2$	$\alpha = 2.4$	$\alpha = 2.6$
ν_μ	0.52	0.46	0.40	0.36
$\overline{\nu}_\mu$	0.52	0.46	0.40	0.36
ν_e	0.28	0.25	0.22	0.20
$\overline{\nu}_e$	0.26	0.23	0.21	0.18

These values are close, though not coincident, with those given in [10]. The difference is of the order 10–30% and is maximal for electron antineutrinos. Such a discrepancy is likely to be due to the large value of the pion charge asymmetry adopted in [10], as discussed in [7,9]. Moreover, the results presented in [9] are in good agreement with ours and confirm that the errors due to our poor knowledge of the interaction details are small compared to the precision achievable by the measurements and to the errors introduced by the approximations made in the calculations.

3.1. *The effect of Neutrino oscillation*

As well established by several experiments, neutrinos undergo flavor oscillations during their propagation in vacuum. Since the typical SNR distance from the Earth is much larger than the neutrino oscillation length around the TeV, the phase of oscillations is very large so that we can safely deal with averaged vacuum oscillations. In this case the flavor oscillation probability can be written as

$$P_{ll'} = \sum_j |U_{lj}^2| \cdot |U_{l'j}^2| \,, \tag{6}$$

where $U_{ll'}$ is the unitary mixing matrix among the lepton flavours $l, l' = e, \mu, \tau$. After the propagation the neutrino flux is

$$F_i = \sum_{j=e,\mu,\tau} P_{ij} F_j^0 \,, \tag{7}$$

where F_j^0 is the expected neutrino flux in the absence of oscillations.

The oscillation probability matrix P_{ij}, calculated in [7] adopting the standard parametrization by [16] and neglecting the unknown CP violating phase δ_{CP}, is used to calculate the ν/γ ratios after propagation which we show in Tab. 2. The flavor composition is almost isotropized by the effect

Table 2. ν/γ ratios at Earth after propagation.

Q_{ν_i}/Q_γ	$\alpha = 2.0$	$\alpha = 2.2$	$\alpha = 2.4$	$\alpha = 2.6$
ν_μ	0.27	0.23	0.21	0.18
$\overline{\nu}_\mu$	0.26	0.23	0.20	0.18
ν_e	0.27	0.24	0.22	0.19
$\overline{\nu}_e$	0.26	0.23	0.20	0.18
ν_τ	0.27	0.23	0.21	0.18
$\overline{\nu}_\tau$	0.26	0.23	0.20	0.18

of the oscillations: the expected fluxes for each flavor are the same within a 20%. Another remarkable effect is the appearing of a ν_τ component, even if it is not generated at the source.

4. Neutrinos from the Galactic Centre

We will apply here our previous general results to estimate the neutrino flux from the dense molecular cloud complex in the Galactic Centre region. A discovery of an extended very-high-energy γ-ray emission coming from a region roughly delimited by $|l| < 0.8°$ and $|b| < 0.3°$ in galactic coordinates has been reported by HESS.[5] The energy spectrum of this source was measured to be $(1.73 \pm 0.13 \pm 0.35) \times E_{\rm TeV}^{-(2.29\pm0.07\pm0.02)}$ TeV^{-1}cm^{-2}s^{-1}sr^{-1}. Superimposed to that diffused emission HESS found a point-like source (J1745-290), which was also earlier detected by HESS itself[17] (and confirmed by MAGIC[18]) with an energy spectrum $(2.50 \pm 0.2) \times E_{\rm TeV}^{-(2.21\pm0.09)}$ TeV^{-1}cm^{-2}s^{-1}, in a position compatible either with the supermassive black hole in Sgr A* or with the SNR Sgr A East. A more recent analysis suggests that the energy spectrum of the point-like source could be a power-law between 125 GeV and 20 TeV with a spectral index $\Gamma = -(2.29 \pm 0.05 \pm 0.10)$ and an integrated flux above 1 TeV of $(1.8 \pm 0.1 \pm 0.3)\mathrm{cm}^{-2}\mathrm{s}^{-1}$.[19]

The detected γ-rays are likely to be due to the hadronic interaction of locally produced cosmic rays with the dense molecular gas,[5] because of the strong correlation found between the γ-ray and the millimetric emissions from the CS in that region. The coincidence, within errors, between the spectral slope of the GC ridge emission detected by HESS and that of J1745-290, suggests that the primary particles responsible for both signals may have the same origin. Indeed, the total CR energy required to explain both of them is compatible with that typically expected from shock acceleration by the remnant of a single SN. These considerations point to Sgr A East as the most plausible primary source (see [8,24] about the identification of J1745-290 with that SNR).

The detection of neutrinos from this region would further support this appealing scenario. The expected neutrino flux can be easily determined by applying the results of Sec. 3. Due the relatively poor angular resolution (of order $\lesssim 1°$) achievable with Neutrino Telescopes and the very low expected neutrino fluxes, it will be quite hard, if possible, to disentangle the neutrino point-like emission from that coming from the Galactic Centre ridge. Thus, we have to sum the neutrino flux from both sources. We use the central value of the slope measured by HESS for the GC ridge, $\alpha = 2.29$. Hence, the total γ-ray flux, above the TeV, coming from the GC region is roughly (we only write central values in the following)

$$F_\gamma^{\rm GC}(E_\gamma) \simeq 7.5 \times 10^{-12} \left(\frac{E_\gamma}{1\ \mathrm{TeV}}\right)^{-2.3} \mathrm{TeV}^{-1}\,\mathrm{cm}^{-2}\,\mathrm{s}^{-1}\ . \tag{8}$$

Using our results of Sec.3, we derive the ν/γ ratios for $\alpha = 2.29$ at the source and at the Earth are shown in Tab. 3 for the different neutrino species [b].

The total expected muon neutrino+antineutrino flux is

$$F_{\nu_\mu+\overline{\nu}_\mu}^{\rm GC}(E_\nu) \simeq 3.3 \times 10^{-12} \left(\frac{E_\nu}{\mathrm{TeV}}\right)^{-2.3} \mathrm{TeV}^{-1}\,\mathrm{cm}^{-2}\,\mathrm{s}^{-1}\ . \tag{9}$$

Using this flux and the ANTARES effective area for up-going ν_μs given in [25], we estimated the expected detection rate by this experiment to be

[b] In [7] we also estimated the contribution to the antineutrino flux due to the decay of high energy neutrons produced by nuclei photodisintegration and *pp* interactions. This was motivated by the possible connection with UHECR anisotropies found by some experiments[20,21] but not confirmed by AUGER.[22] Unfortunately, also the antineutrino flux happens to be very small, compared to that expected by pion decay.[7]

Table 3. ν/γ ratios before and after oscillations.

$\alpha = 2.29$	at source	at Earth
ν_e	0.24	0.23
$\overline{\nu}_e$	0.21	0.21
ν_μ	0.43	0.22
$\overline{\nu}_\mu$	0.43	0.22
ν_τ	0	0.22
$\overline{\nu}_\tau$	0	0.22

0.07 yr^{-1} which is, unfortunately, quite small. A km^3 Neutrino Telescope placed in the Mediterranean sea, e.g. NEMO[26] or NESTOR,[27] would have better chances to find a positive result because of its effective area about 20 times larger than that of ANTARES. The expected rate we estimate for such a detector is 1.5 yr^{-1}. Since the estimated background rate of such device is ~ 0.3 ev/yr in a $(1°)^2$ angular bin, the observation time required to detect that source at 95%C.L. is about three years.

5. Conclusions

We estimated the neutrino/photon flux ratio in the TeV energy range from *embedded SNRs*, under the hypothesis that SNRs shock accelerate protons with a power-law energy spectrum at least up to the PeV. We computed first the relative emissivities of muon and electron neutrinos produced by charged pion decay. Then, we accounted for vacuum oscillations to determine the ν_i/γ flux ratio expected on the Earth for every lepton family. Our results are similar to those derived by other authors.[9,10,23]

As an application of our results we used the data concerning the GC region obtained by the HESS collaboration[5] to estimate the muon neutrino flux from that region. We found that a km^3 Neutrino Telescope in the Mediterranean sea may be able to detect that source (see also [12]). The possible observation of a positional coincidence between the γ-ray and the neutrino source and the measurement of a flux ratio at the level estimated here, would add strong evidence favouring the hypothesis that hadron acceleration is taking place in that region. Furthermore, since embedded SNRs should be quite common objects in the Galaxy and VHE γ-ray and neutrino astronomy has just started, other sources of that kind may be discovered in both channels in a not too far future.

References

1. T. Montmerle, Astrophys. J. **231** (1979) 95.
2. F.A. Aharonian and A.M. Atoyan, Astron. Astrophys. **309** (1996) 917.
3. S.D. Hunter *et al.*, Astrophys. J. **481** (1997) 205.
4. F. Aharonian *et al.* [The HESS Collaboration], Nature, **432** (2004) 75.
5. F. A. Aharonian [HESS Collaboration], Nature **439** (2006) 695, [arXiv:astro-ph/0603021].
6. T. Montaruli, Eur. Phys. J. A **24S1** (2005) 103.
7. V. Cavasinni, D. Grasso and L. Maccione, Astropart. Phys. **26** (2006) 41 [arXiv:astro-ph/0604004].
8. D. Grasso and L. Maccione, "Sgr A East as a possible high energy neutron factory in the galactic Astropart. Phys. **24** (2005) 273 [arXiv:astro-ph/0504323].
9. S. R. Kelner, F. A. Aharonian and V. V. Bugayov, "Energy spectra of gamma-rays, electrons and neutrinos produced at proton Phys. Rev. D **74** (2006) 034018 [arXiv:astro-ph/0606058].
10. M. L. Costantini and F. Vissani, Astropart. Phys. **23** (2005) 477 [arXiv:astro-ph/0411761].
11. M. L. Costantini and F. Vissani, AIP Conf. Proc. **794** (2005) 219 [arXiv:astro-ph/0508152].
12. M. D. Kistler and J. F. Beacom, arXiv:astro-ph/0607082.
13. P. Blasi and S. Colafrancesco, Astropart. Phys. **122** (1999) 169 [arXiv:astro-ph/9905122].
14. T. Sjostrand, L. Lonnblad, S. Mrenna and P. Skands, arXiv:hep-ph/0308153.
15. T.K. Gaisser, Cosmic Rays and Particle Physics, Cambridge University Press, 1990.
16. S. Eidelman *et al.* [Particle Data Group], Phys. Lett. B **592** (2004) 1.
17. F. Aharonian *et al.* [The HESS Collaboration], Astron. Astrophys. **425** (2004) L13 [arXiv:astro-ph/0408145].
18. J. Albert *et al.* [MAGIC Collaboration], Astrophys. J. **638** (2006) L101 [arXiv:astro-ph/0512469].
19. L. Rolland and J. Hinton for the HESS Collaboration, Proceedings of the 29th ICRC, Pune (2005) **4**, 109-112.
20. D. J. Bird *et al.* [HIRES Collaboration], arXiv:astro-ph/9806096.
21. N. Hayashida *et al.* [AGASA Collaboration], Astropart. Phys. **10** (1999) 303 [arXiv:astro-ph/9807045].
22. A. Letessier-Selvon [Pierre Auger Collaboration], 29th International Cosmic Ray Conference, Pune, 2005, [arXiv:astro-ph/0507331].
23. R. M. Crocker, F. Melia and R. R. Volkas, Astrophys. J. **622** (2005) L37 [arXiv:astro-ph/0411471].
24. R. M. Crocker, M. Fatuzzo, R. Jokipii, F. Melia and R. R. Volkas, Astrophys. J. **622** (2005) 892 [arXiv:astro-ph/0408183].
25. E. Aslanides *et al.* [ANTARES Collaboration], arXiv:astro-ph/9907432.
26. C.N. De Marzo [NEMO Collaboration], Nucl, Phys. *Proc. Suppl.* **B100** (2001) 344.
27. S. E. Tzamarias [NESTOR Collaboration], Nucl. Instrum. Meth. A **502** (2003) 150.

Solving GRBs and SGRs puzzles by precessing Jets

D.Fargion, O.Lanciano, P.Oliva

Physics Department, University of Rome, "La Sapienza", and INFN, Ple.A.Moro 5, 00185,Rome, Italy
** E-mail: ab_daniele.fargion@roma1.infn.it*

A persistent, thin (a solid angle less than 0.1μsr.), gamma jet, may be ejected from BH and Pulsars, powered by ultra-relativistic electron pairs. These γ jet while precessing and spinning are originated by Inverse Compton[7] and-or Synchrotron Radiation at pulsars or micro-quasars sources. They are most powerful at Supernova birth blazing, once on axis, to us and flashing GRB detector. The trembling of the thin jet explains naturally the observed erratic multi-explosive structure of different GRBs. The jets are precessing (by binary companion) and decaying on time scales of a few hours but they are usually staying inside the observer cone view only a few seconds (GRB) duration times; the jets whole lifetime, while decaying in output, could survive as long as thousands of years, linking huge GRB-SN jet apparent Luminosity to more modest SGR relic Jets (at corresponding X-Ray pulsar output). Therefore long-life SGR may be repeating and if they are around our galaxy they might be observed again as the few known ones and a few rarer extragalactic XRFs. The orientation of the beam respect to the line of sight plays a key role in differentiating the wide GRB morphology. The relativistic cone is as small as the inverse of the electron progenitor Lorentz factor ($\gamma_e \simeq 10^3 - 10^4$ for I.C.[7] and $\gamma_e \simeq 10^8 - 10^9$ for Synchrotron radiation). The hardest and brightest gamma spectra are hidden inside the inner gamma jet axis. To observe the inner beamed GRB events one needs the widest SN sample and largest cosmic volumes (red-shift ≥ 1). The most beamed are hardest. On the contrary the nearest ones, within tens Mpc distances, are mostly observable on cone periphery. Their consequent large impact crossing angle, leads to longest *anomalous* SN-GRB duration, with lowest fluency and the softest spectra, as in GRB060218 signature. A majority of GRB jet blazing much later (weeks, months after) may hide their progenitor explosive SN after-glow and therefore are so called *orphan* GRB. The late GRBs are called now local SGRs (or in outer extragalactic XRF) and are linked to anomalous X-ray AXPs. Conical shape of few nebulae and the precessing jet of few known micro-quasar describe in space the model $3D$ signature. Recent outstanding episode of X-ray precursor, ten minutes before the main GRB event, cannot be understood otherwise.

Keywords: Gamma Ray Burst, Soft Gamma Repeaters, Inverse Compton Scattering, Muons

1. Introduction: GRB-SGR open questions

Why GRBs are so spread in their total energy, (above 6 orders of magnitude) and in their peak energy following the so-called Amati correlation[1]? Does the Amati law imply more and more new GRB families? Why, as shown below the GRB energy is not a constant but a growing function (almost quadratic) of the red-shift? Why are the harder and more variable

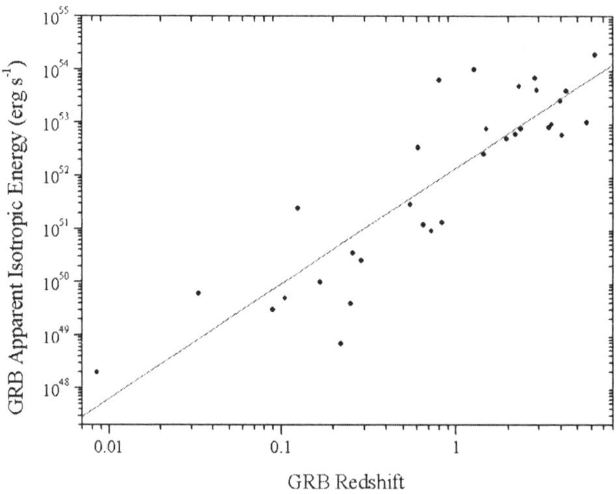

Fig. 1. Most Recent Swift GRB isotropic apparent energy[1] versus their observed red-shift in Log-Log plot: the nearest GRB980425, $z = 0.008$, as well as GRB020903, GRB030329, GRB031203 and most distant GRB one at $z \simeq 6$ have been added and they appear at the line extreme. The consequent most correlated law for the apparent energy growth with distances is nearly quadratic in the redhift: $E_i so = 10^{52} \cdot z^2$ erg. Therefore GRB are apparently *not standard candle* but growing in power *quadratically with their red-shift.* This puzzle may be solved assuming both threshold cut off for far off-axis GRB and extreme beaming at cosmic edges

GRBs (,[38]) found at higher redshifts contrary to expected Hubble law? Why does the output power of GRB vary in a range (see also[8]) of $8-9$ orders of magnitudes with the most powerful events residing at the cosmic edges ([19])? Why has it been possible to find in the local universe(at distances $40-150 Mpc$ just a part over a million of cosmic space) at least two nearby events (GRB980425 at z=0.008 and recent GRB060218 at z=0.03). Most GRBs should be located at largest volumes, at $z \geq 1$, ([8]). Why are these two nearby GRBs so much under-luminous ([8])? Why are their time evolution so slow and smooth? Why do their afterglows show so many bumps and re-brightening as the well-known third nearest event, GRB030329? Why do not many GRB curves show monotonic decay (an obvious consequence of a one-shot explosive event), rather they often show sudden re-brightening or bumpy afterglows at different time scales and wavelengths (;[17];[10] see e.g. GRB 050502B,[5]) ? Why have there been a few GRBs and SGRs whose spectra and time structure are almost identical if their origin is so different (beamed explosion for GRB versus isotropic magnetar) (,[818])? How can a jetted fireball (with an opening angle of $5^o - 10^o$) release a power nearly 6 orders of magnitude more energetic than the corresponding isotropic SN? How can re-brightening take place in the X-ray and optical afterglows ([10])? How can some ($\sim$ 6%) of the GRBs (or a few SGRs) survive the 'tiny' (but still extremely powerful) explosion of its *precursor* without any consequences, and then explode, catastrophically, a few minutes later? In such a scenario, how could the very recent GRB 060124 (at redshift z=2.3) be preceded by a 10 minutes precursor, and then being able to produce multiple bursts hundreds of times brighter? Why SGR1806-20 on 27th Dec. 2004 , show no evidence of the loss of its period P or its derivative $\dot{P}$ after the huge *Magnetar* eruption while in this model its hypothetical magnetic energy reservoir (linearly proportional to $P \cdot \dot{P}$) must be largely exhausted? Why do SGR1806 radio afterglows show a mysterious two-bump radio curve implying additional energy injection many days later? In this connection why are the GRB021004 light curves (from X to radio) calling for an early and late energy injection? Why has the SGR1806-20 polarization curve been changing angle radically in short ($\sim$ days) timescale? Why is the short GRB050724 able to bump and re-bright a day after the main burst[2]? Once these major questions are addressed and (in our opinion) mostly solved by our precessing gamma jet model, a final question still remains, calling for a radical assumption on the thin precessing gamma jet: how can an ultra-relativistic electron beam (in any kind of Jet models) survive the SN background and dense matter layers and escape in the outer space while remaining collimated?

Such questions are ignored in most Fireball models that try to fit the very different GRB afterglow light curves with shock waves on tuned shells and polynomial ad-hoc curves around the GRB event. Their solution forces us more and more toward a unified precessing Gamma Jet model feeded by the PeV-TeV lepton showering (about UHE showering beam see analogous ones[9]) into γ discussed below. As we will show, the thin gamma precessing jet is indeed made by a chain of primary processes (PeV muon pair bundles decaying into electrons and then radiating via synchrotron radiation), requiring an inner ultra-relativistic jet inside the source.

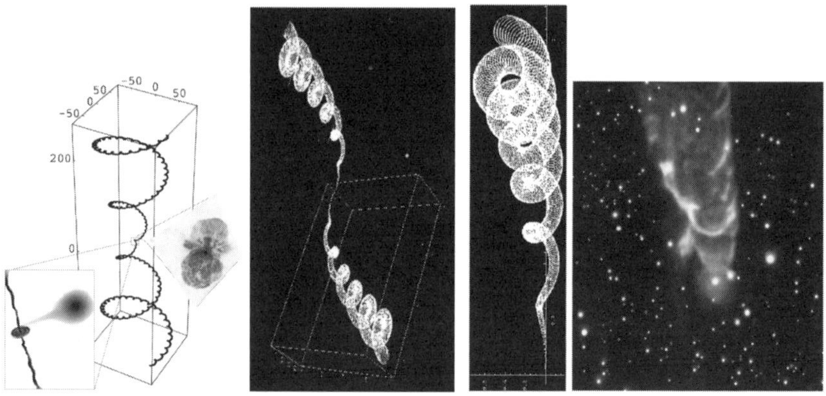

Fig. 2. From the left to the right: A possible 3D structure view of the precessing jet obtained with, for instance, a non linear precessing, while spinning, gamma jet; at its center the "explosive" SN-like event for a GRB or a steady binary system for a SGRs where an accretion disc around a compact object powers a collimated precessing jet. In the two center figures the 3 D and the projected 2D of such similar precessing Jet. In the right last panel we show an Herbig Haro - like object HH49, whose spiral jets are describing, at a lower energy scale, the ones in micro-quasars such as well known SS-433.

2. Blazing Precessing jets in GRBs and SGRs

The huge GRBs luminosity (up to 10^{54} erg s^{-1}) may be due to a high collimated, on-axis blazing jet powered by a Supernova output; the gamma jet is made by relativistic synchrotron radiation (or ICS) and the inner the jet the harder and the denser is its output. The harder the photon energy, the thinner is the jet opening angle $\Delta\theta \simeq \gamma^{-1}$, $\Delta\Omega \simeq \gamma^{-2}$, $\gamma \simeq 10^4$. The thin solid angle explains the rare SN-GRB connection and for instance the apparent GRB 990123 extraordinary power (billions of times the typical SN luminosity). This also explains the rarer, because nearer, GRB-SN events such as GRB980425 or GRB060218, whose jets were off-axis $\simeq 300 \cdot \gamma^{-1}$, i.e. a few degrees, increasing its detection probability by roughly a hundred thousand times , but whose GRB luminosity was almost by the same factor extremely low. This beaming selection in larger volumes explains

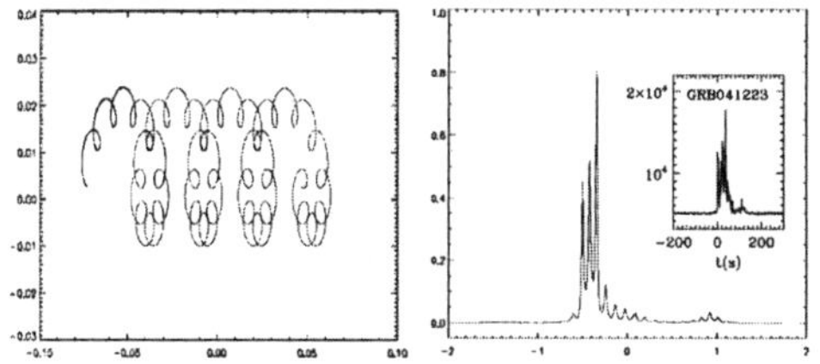

Fig. 3. The possible simple beam track of a precessing jet to observer located at origin. On the left the observer stay at 0.00 – 0.00; the progenitor electron pair jet (leading by I.C. citeFaSa98 to a gamma jet) has here a Lorentz factor of a thousand and consequent solid angle at μsr. Its consequent blazing light curve corresponding to such a similar outcome observed in GRB041223.

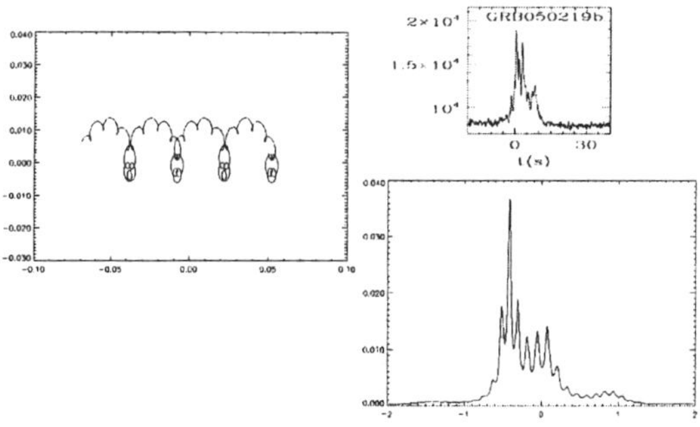

Fig. 4. As above a Precessing jet and its consequent light curve versus a similar outcome observed in GRB050219b.

the puzzling evidence (the Amati correlation) of harder and apparently powerful GRBs at larger and larger distances (in opposition to Hubble law calling for redder and redder signals). The statistical selection favors (in wider volumes and for a wider sample of SN-GRB-jet) the harder and more on-axis events. A huge (million time) unobserved population of far off-axis SN-GRBs at cosmic distances are below the detection thresholds. This Amati correlation remains unexplained for any isotropic or fountain

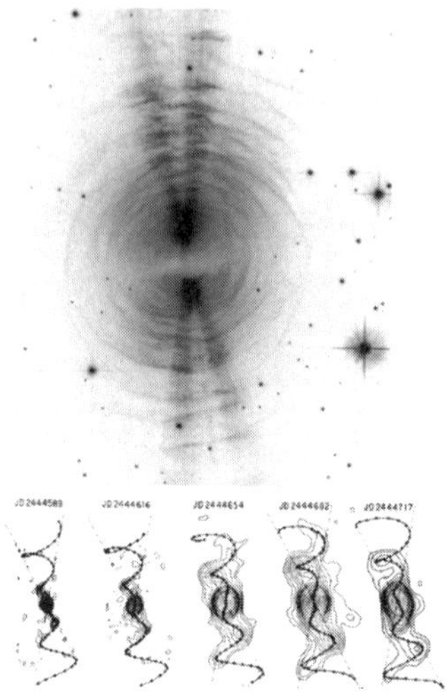

Fig. 5. The Egg Nebula, whose shape might be explained as the conical section of a twin precessing jet interacting with the surrounding cloud of ejected gas. *Down:* The similar observed structure of the outflows from the microquasar SS433. A kinematic model of the time evolution of two oppositely directed precessing jets is overlaid on the radio contours (from Blundell & Bowler 2005).

cone Fireball model and it is in contrast with the cosmic trend required by the Hubble-Friedmann law: the further the distances, the larger the redshifts, the smoother the time lag profile and the softer the expected GRB event. In our model to make GRB-SN in nearly energy equipartition the jet must be very collimated $\frac{\Omega}{\Delta\Omega} \simeq 10^8 - 10^{10}$ (Fargion,Salis 1995;Fargion 1999; Fargion,Grossi 2005) explaining why apparent (but beamed) GRB luminosity $\dot{E}_{GR-jet} \simeq 10^{53} - 10^{54} ergs^{-1}$ coexist on the same place and similar epochs with lower (isotropic) SN powers $\dot{E}_{SN} \simeq 10^{44} - 10^{45} ergs^{-1}$. In order to fit the statistics between GRB-SN rates, the jet must have a decaying activity ($\dot{L} \simeq (\frac{t}{t_o})^{-\alpha}$, $\alpha \simeq 1$), it must survive not just for the observed GRB duration, but for a much longer timescale, possibly thousands of time longer, $t_o \simeq 10^4 s$. The late stages of the GRBs (within the same decaying power law) would appear as a SGRs. Indeed the puzzle (for one shot popular Magnetar-Fireball model) arises for the surprising giant flare from SGR 1806-20 that occurred on 2004 December 27: if it has been radiated isotropically (as assumed by the Magnetar model), most of (if not all) the magnetic energy stored in the neutron star NS should have been consumed at once. This should have been reflected into sudden angular ve-

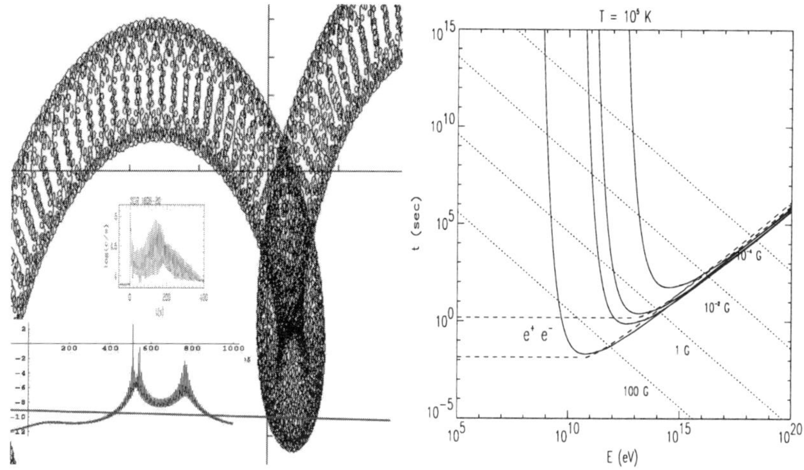

Fig. 6. Left: The more structured multi precessing jet able to describe the most rare blazing and oscillating SGR 1806-20. The inner spirals are reflecting the precursor trembling while the Jet lorentz factor is assumed here at $\gamma_e \simeq 10^9$, but in general the jet showering structure is not cone-like but fan-like because external stellar magnetic field presence[9] where $\theta \simeq {\gamma_e}^{-1}$ Right:The electron and muon interaction lengths. The *dashed-dotted* and *dotted* lines correspond to the synchrotron energy loss distance (for muons and electrons respectively) for different values of the magnetic field: 100 G, 1 G and 10^{-2} G. The *straight solid* line labelled t_μ indicates the muon lifetime; the *dashed* lines indicate the IC interaction lengths for muons and electrons. Finally the two *solid* curves labelled $\mu^+\mu^-$ and e^+e^- correspond to the attenuation length of high energy photons producing lepton pairs (either $\mu^\pm$ or $e^\pm$) through the interaction with the SN radiation field. We have assumed that the thermal photons emitted by the star in a pre-SN phase have a black body distribution with a temperature $T \simeq 10^5$ K. Assuming a radius $R \sim 10\,R_\odot$, we are considering a luminosity of $L_{SN} \simeq 2.5 \cdot 10^{41}$ erg s^{-1}. Around $10^{15} - 10^{16}$ eV muons decay before losing energy via IC scattering with the stellar background or via synchrotron radiation. Right :The Supernova opacity (interaction length) for PeV electrons at different times. PeV muon jets may overcome it and decay later in γ showering electrons (see for details Fargion, Grossi 2005);

locity loss (and-or its derivative) never observed. On the contrary a thin collimated precessing jet $\dot{E}_{SGR-jet} \simeq 10^{36} - 10^{38} ergs^{-1}$, blazing on-axis, may be the source of such an apparently (the inverse of the solid beam angle $\frac{\Omega}{\Delta\Omega} \simeq 10^8 - 10^9$) huge bursts $\dot{E}_{SGR-Flare} \simeq 10^{38} \cdot \frac{\Omega}{\Delta\Omega} \simeq 10^{47} ergs^{-1}$ with a moderate steady jet output power (X-Pulsar, SS433). This explains the absence of any variation in the SGR1806-20 period and its time derivative, contrary to any obvious correlation with the dipole energy loss law. The nearby spiralling of the jet to us explains the later bumping oscillatory signal of the event (see figure below).

In our model, the temporal evolution of the angle between the jet di-

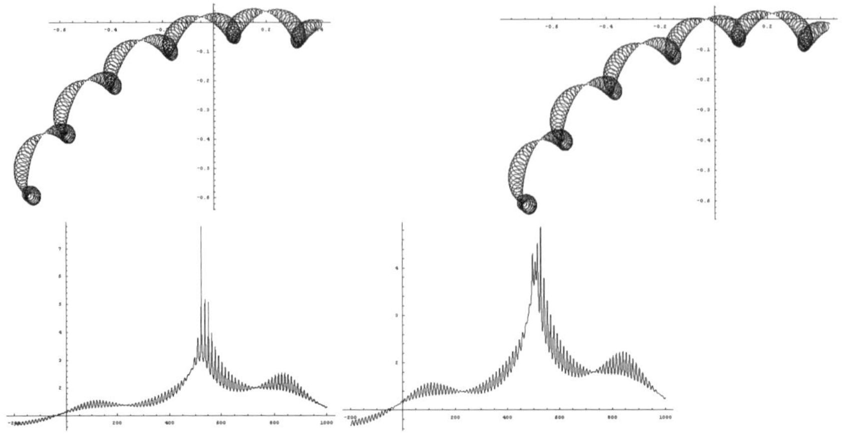

Fig. 7. Left and right a more beamed or less beamed (non linear) precessing blazing: note that a much off-axis beaming induce a different SGR smoother and softer profile and a much limited GRB amplification

rection and the rotational axis of the NS can be expressed as

$$\theta_1(t) = \sqrt{\theta_x^2 + \theta_y^2}$$

, where

$$\theta_y(t) = \theta_a \cdot \sin \omega_0 t + \cos(\omega_b t + \phi_b) +$$

$$\theta_{psr} \cdot \cos(\omega_{psr} t + \phi_{psr}) \cdot |(\sin(\omega_N t + \phi_N))| + \theta_s \cdot \cos(\omega_s t + \phi_s) + \theta_N \cdot \cos(\omega_N t + \phi_N)) + \theta_y(0)$$

A similar law express the $\theta_x(t)$ evolution. (where γ is the Lorentz factor of the relativistic particles of the jet, see Table I for the most powerful SGR1806-20 event. See,[11,12]

$\gamma = 10^9$	$\theta_a = 0.2$	$\omega_a = 1.6 \cdot 10^{-8}$ rad/s
$\theta_b = 1$	θ_{psr}=1.5 $\cdot 10^7/\gamma$	θ_N=5 $\cdot 10^7/\gamma$
ω_b=4.9 $\cdot 10^{-4}$ rad/s	ω_{psr}=0.83 rad/s	ω_N=1.38 $\cdot 10^{-2}$ rad/s
$\phi_b = 2\pi - 0.44$	$\phi_{psr} = \pi + \pi/4$	ϕ_N=3.5 $\pi/2 + \pi/3$
$\phi_s \sim \phi_{psr}$	θ_s=1.5 $\cdot 10^6/\gamma$	$\omega_s = 25$ rad/s

The simplest way to produce the γ emission would be by IC of GeVs electron pairs onto thermal infra-red photons. Also electromagnetic showering of PeV electron pairs by synchrotron emission in galactic fields, ($e^{\pm}$

from muon decay) may be the progenitor of the γ blazing jet. However, the main difficulty for a jet of GeV electrons is that their propagation through the SN radiation field is highly suppressed. UHE muons ($E_\mu \gtrsim$ PeV) instead are characterized by a longer interaction length either with the circum-stellar matter and the radiation field, thus they have the advantage to avoid the opacity of the star and escape the dense GRB-SN isotropic radiation field,[11] DaF06. We propose that also the emission of SGRs is due to a primary hadronic jet producing ultra relativistic $e^{\pm}$ (1 - 10 PeV) from hundreds PeV pions, $\pi \to \mu \to e$, (as well as EeV neutron decay in flight): primary protons can be accelerated by the large magnetic field of the NS up to EeV energy. The protons could in principle emit directly soft gamma rays via synchrotron radiation with the galactic magnetic field ($E^p_\gamma \simeq 10(E_p/EeV)^2(B/2.5 \cdot 10^{-6}\,G)$ keV), but the efficiency is poor because of the too small proton cross-section, too long timescale of proton synchrotron interactions. By interacting with the local galactic magnetic field relativistic pair electrons lose energy via synchrotron radiation, $E^{sync}_\gamma \simeq 4.2 \times 10^6 \left(\frac{E_e}{5\cdot 10^{15}\,eV}\right)^2 \left(\frac{B}{2.5\cdot 10^{-6}\,G}\right)\, eV$, with a characteristic timescale $t^{sync} \simeq 1.3 \times 10^{10} \left(\frac{E_e}{5\cdot 10^{15} eV}\right)^{-1} \left(\frac{B}{2.5\cdot 10^{-6}\,G}\right)^{-2}\, s$. This mechanism would produce a few hundreds keV radiation as it is observed in the intense γ-ray flare from SGR 1806-20. The Larmor radius is about two orders of magnitude smaller than the synchrotron interaction length and this may imply that the aperture of the showering jet is spread in a fan structure[9] by the magnetic field, $\frac{R_L}{c} \simeq 4.1 \times 10^8 \left(\frac{E_e}{5\cdot 10^{15} eV}\right) \left(\frac{B}{2.5\cdot 10^{-6}\,G}\right)^{-1}\, s$. Therefore the solid angle is here the inverse of the Lorentz factor (nano-sr.)In particular a thin ($\Delta\Omega \simeq 10^{-9} - 10^{-10}\, sr$) precessing jet from a pulsar may naturally explain the negligible variation of the spin frequency $\nu = 1/P$ after the giant flare ($\Delta\nu < 10^{-5}$ Hz). Indeed it seems quite unlucky that a huge ($E_{Flare} \simeq 5 \cdot 10^{46} erg$) explosive event (as the needed mini-fireball by a magnetar model; Duncan et al. 1992) is not leaving any trace in the rotational energy of the SGR 1806-20, $E_{rot} = \frac{1}{2} I_{NS}\omega^2 \simeq 3.6 \cdot 10^{44} \frac{P}{7.5s}^{-2} \left(\frac{I_{NS}}{10^{45} g\, cm^2}\right) erg$. The consequent fraction of energy lost after the flare is severely bounded by observations : $\frac{\Delta(E_{Rot})}{E_{Flare}} \leq 10^{-6}$. The thin precessing Jet while being extremely collimated (solid angle $\frac{\Omega}{\Delta\Omega} \simeq 10^8 - 10^{10}$ (Fargion,Salis 1995;Fargion 1999;[11]);[12]) may blaze at different angles within a wide energy range (inverse of $\frac{\Omega}{\Delta\Omega} \simeq 10^8 - 10^{10}$). The output power may exceed $\simeq 10^8$, explaining the extreme low observed output in GRB980425, an off-axis event, the long late off-axis gamma tail by GRB060218,[13]), respect to the on-axis and more distant GRB990123 (as well as GRB050904). **In conclusion *GRBs are***

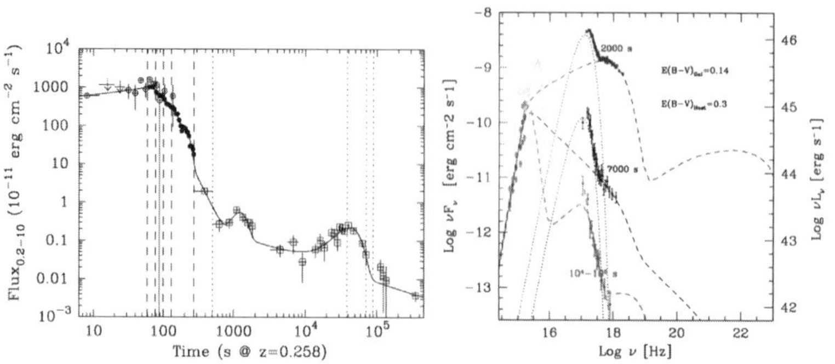

Fig. 8. Left: The short GRB050724 and its long life X-ray afterglow whose curve ([2]) and whose multi-re-brightening is testing the persistent jet activity and geometrical blazing views. Right: the very recent optical afterglows of GRB 060218 whose smooth longest X-ray flare coexist with a thermal Black Body Radiation (see previous figure) component (a steady SN bump) while the external Jet cone tail are fading while pointing off-axis elsewhere, adapted from.[14] (;[16,13]).

not the most powerful explosions, but just the most collimated ones.

References

1. Amati L., Frontera F.et al., A&A, 390 (2002) 81 and astro-ph/0611189
2. Campana, S., et. al. astro-ph/0603475, A&A accepted
3. Lazzati D. et al. astro-ph/0602216
4. Duncan, R., & Thompson, C., 1992, ApJ, 392, L9
5. Falcone A. et al. astro-ph/0512615 , astro-ph/0602135
6. Fargion, D., Salis, A., 1995b, Astrophysics & Space Science, 231, 191
7. Fargion, D., Salis, A., 1998, Phys.Usp. 41 (1998) 823-829astro-ph/9605168
8. Fargion,D., 1999, AA&SS, 138, 507; astro-ph/9903433, astro-ph/9808005
9. D.Fargion,B.Mele, A.Salis, Astrophys.J. 517 (1999) 725-733; D.Fargion,Astrophys.J. 570 (2002) 909-925; D.Fargion et al.Astrophys.J. 613 (2004) 1285-1301
10. Fargion, D, Chin.J.Astron.Astrophys. 3 (2003) 472-482
11. Fargion, D.,Grossi M.,astro-ph/0504638, Nuovo Cim.28C(2005)809-812
12. Fargion, D., Grossi, M.,Chin.J.Astron.Astrophys. 6S1 (2006) 342-348
13. Fargion, D. GNC 4819, 06/02/23
14. G. Ghisellini, G. Ghirlanda, F. Tavecchio, astro-ph/0608555
15. Medina-Tanco, G.A., Watson, A.A, 2001, *Proceedings of ICRC*, 531.
16. Moretti, A., 2006, *www.merate.mi.astro.it/ moretti/lc*$_0$*60218.gif*
17. Stanek, R., et al.astro-ph/0602495
18. Woods, P.M., et al. 1999, ApJ,527,L47; astro/ph 9909276
19. Yonetoku, D., Murakami, T., Nakamura, T., 2004, ApJ, 609, 935

IV Extragalactic Sources

MULTIWAVELENGTH OBSERVATIONS AND THEORIES OF BLAZARS

G. TOSTI

Department of Physics & INFN Perugia
Via A. Pascoli, Perugia, 06100 Perugia, Italy
**E-mail: gino.tosti@pg.infn.it*

Multiwavelength (MW) observations are an essential diagnostic tool to study the physics of blazars. Blazars, the most extreme objects among the Active Galactic Nuclei (AGNs), are characterized by rapid variability at all wavelengths from radio through TeV γ-rays. Considerable progress has been made in recent years thanks to multiwavelength monitoring campaigns. In this contribution, I will give a brief overview of the MW campaigns carried out in the past by large collaboration and their impact on the knowledge of the most important source/jet parameters needed by the physical models used to fit the observed blazar spectral energy distribution (SED). Furthermore,I will discuss some of the perspectives for multiwavelength observations during the operation of the gamma-ray mission GLAST.

Keywords: AGN; Blazars; Multiwavelength observations; Emission models

1. Introduction

Blazars are rare and extreme objects and represent only a small fraction of radio-loud AGNs (about the 15% – 20% of the total), characterized broad, non-thermal, polarized, and highly variable continuum flux, extending over the whole electromagnetic spectrum. According to the presence or the absence ($EW < 5\AA$) of broad emission lines, in the optical spectrum, they are classified as Flat Spectrum Radio Quasars (FSRQ) or as BL Lac objects respectively. Radio observations of blazar have been important to reveal the polarization of the radiation, the morphology of the host galaxy extended emission and the presence of a relativistic jet where dicrete knots are moving with apparent superluminal speeds (see e.g. Ref. 10).

The Energetic Gamma Ray Experiment Telescope (EGRET) onboard the Compton Gamma Ray Observatory (CGRO) discovered about 70 blazars emitting γ-rays (E 30 MeV, 3rd EGRET Catalog[6]). In the radio band the γ-ray sources detected by EGRET have VLBI components having faster speeds than the non-EGRET sources.[9,10,13]

About one dozen blazars, and in particular BL Lac objects, have also been detected by Ground-based Cerenkov telescopes at E >100 GeV. TeV Blazars are highly variable in both the TeV-ray and X-ray bands. In the radio band they are characterized by weak flux and low levels of variability and subluminal VLBI pattern speed.[18]

Thanks to EGRET and Cerenkov telescopes we now know the overall spectral energy distribution (SED) of blazars.[24] It is characterized by two pronounced components separated by 8-10 decades in frequency[22] (see Fig. 1). The lower frequency component has peak frequency in the IR/X-ray energy range, the higher frequency component has peak frequency in the MeV-GeV energy range. Based on this characteristic of the SED, the BL Lacs objects have been divided[16] in High-energy peaked BL Lacs (HBL, objects which have the low energy peak in the UV/softX range) and Low-energy peaked BL Lacs (LBL , objects which have the low energy peak in the far-IR/near-IR).

From the analysis of the SEDs (from radio to γ) of large sample of blazars have been proposed that all blazars lies on a sequence governed by a single physical parameter: the bolometric luminosity.[5] This picture is, however, no free of controversy (see Ref.17 for a recent review)

According to the AGNs unified schemes,[23] blazars are radio galaxies with their radio jets forming a small angle with respect to the line of sight (Blandford & Rees 1978) . Probably, their nonthermal continuum emission is produced in sub-parsec sized emission regions, (by ultrarelativistic particles) moving at relativistic speeds corresponding to bulk Lorentz factors Γ around 10. The radiation emitted by these regions is Doppler boosted (beamed radiation) into the direction of the observer and dominates over other radiative components of the AGN (unbeamed radiation) The source of power for the relativistic jet is the accretion of matter (from the host galaxy) onto a supermassive black hole.

In the framework of relativistic jet models the low energy component of the blazar's SED is produced through synchrotron emission. The high-energy (X-ray to γ-ray) component is thought to be produced via Compton upscattering of low-frequency radiation by the same electrons responsible for the synchrotron emission.

The unified model for radioloud AGNs identifies the Fanaroff-Riley II (FRII) radio galaxies with the parent populations (i.e. the misaligned counterparts with respect to the jet direction) of FSRQ and the Fanaroff-Riley

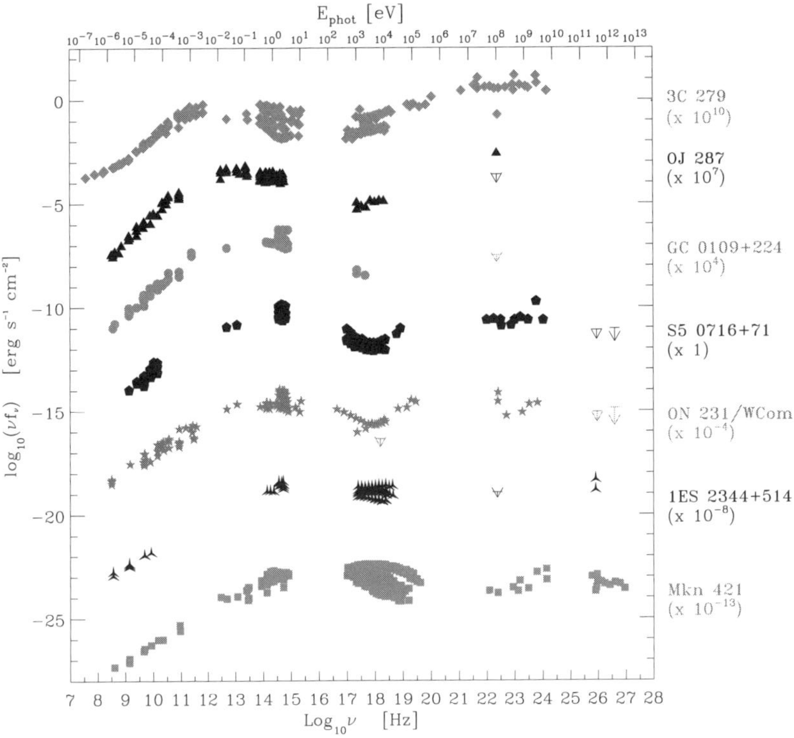

Fig. 1. Typical Spectral Energy Distribution of FSRQs,LBL and HBL

I (FRI)ones, with the parent populations of BL Lac objects. This association is supported by the similar isotropic properties as well as the relative number density of such (parent and beamed) populations.[23]

Recently *Chandra* have discovered intense X-ray emission from large scale jets in both FRI and FRII[19]radio galaxies. The models developed to explain this X-ray emission from large-scale jets, predict also a significant emission in the γ-ray domain detectable by the next-generation of high-energy instruments, such as GLAST.

In the following sections I will give a brief overview of the importance of MW campaigns for the knowledge of the most important source/jet parameters needed by the physical models used to explain the observed blazar spectral energy distribution (SED).

2. MW Observations & Blazars Emission Models

Studies of the properties of blazar radiation as spectrum shape and MW variability provide exceptional tools for exploring the deepest parts of extragalactic jets, their composition, structure, physics, and origin.

The first MW campaigns carried out in 1980's, in optical, UV and X-ray, lead to a scenario in which shocks propagating outward in the jet are the origin of the observed blazar activity.[15] Furthermore the intrinsic difference between the HBL and LBL/FSRQ SEDs was soon recognized.

In 1990's the *EGRET* results, and the detection of the first sources at TeV energies by the *Whipple*, *HEGRA*, *CAT* and *CANGAROO* experiments, have allowed to know the overall spectral energy distribution (SED) of blazars.[24] The γ-ray fluxes show rapid and high-amplitude variability, and in many FSRQ, at least during their high states, they emit most of their energy at these energies.

The mechanism responsible for this high-energy emission is still unclear. While there is large consensus about the synchrotron origin of the low–energy component of the SED, to explain the high-energy emission, two fundamentally different classes of models have been proposed. In *leptonic models*, the high-energy emission is produced via inverse Compton scattering of the same ultrarelativistic electrons producing the synchrotron emission. In this scenario, the principal matter of debate is the origin of the target photons, they may be the same synchrotron photons produced within the jet or external photons (EC process). External photons include accretion-disk photons entering the jet emission region directly or radiation reprocessed in surrounding material like: the broad-line regions; jet synchrotron emission reprocessed by circumnuclear material; infrared emission from circumnuclear dust; synchrotron radiation from other emission regions along the jet. Quite often, a combination of these models is required to fit the SED, especially in the case of FSRQ and LBL blazars. In *hadronic models* the high energy component is explained by: highly relativistic baryonic outflow which sweeps up ambient matter; interactions of high-energy protons with gas clouds moving across the jet; interactions of ultra–high energy protons with ambient photons, with the jet magnetic field or with both. A recent review and original references for all these models can be found in in Ref. 3

Since the models predict different relationships between variations in the different observing bands, intensive simultaneous monitoring in several widely-spaced bands are crucial to get the information on the radiation mechanisms and the structure of the jet. For that reason, several coordinated MW campaigns have been carried out on 3C 279 and a few other EGRET-detected blazars. The 3C 279 SED was measured at different epochs and demonstrates that blazars exhibit not only temporal, but also considerable spectral variability.[7,8] The last MW campaign on 3C 279 have been carried out in January 2006, preliminary results can be found in Ref. 3. However we have not conclusive results about modeling of all available simulataneous MW data on this source.[3]

In addition to planned coordinated campaign, target of opportunity (ToO) observations used optical/X-ray triggers to activate high energy observing programs. Most of the blazars ToO program were triggered by optical observations, and the ToO triggered by the spectacular 1997 flare of BL Lacerte, the prototype of the class of BL Lac objects, can be considered the best example. During this flare ISO, XTE, $ASCA$ and $EGRET$ and many ground based facilities simultaneously observed BL Lacertae.[14]

Other successfully ToO program have been organized during the operation of the X-ray satellites *Beppo*SAX in the 1 – 200 KeV energy range.[21] This is the region of the electromagnetic spectrum where the two broad components of FSRQs and LBLs SED are overlapping and where is the peak of HBLs low energy component. X-ray observations of most of the LBLs observed with *Beppo*SAX at different epochs show significant flux and spectral variability, indicating that the X-ray emission is at times dominated by the high-energy end of the synchrotron emission, while at other occasions it is dominated by the low-frequency portion of the high-energy bump of the SED. These observations have also show strong shifts of the synchrotron peak frequency for Mkn 501 and Mkn 421, two of the most studied TeV blazars, during large flares. Mkn 501 was observed in outburst by BeppoSAX in April 1997 and has revealed an exceptionally hard X-ray spectrum, peaking at or beyond 100 keV (0.3 KeV during quiescence). During this flare the sources have also show a very high flux at TeV energies.[1] The analysis of the simultaneous MW data has permitted to study in details the SED of TeV blazars, and it seem that a relatively simple one/two-components SSC model are able to explain the data (see Ref. 11, for recent review of TeV blazar models). However recent observations in Mrk 421[2] and 1ES 1959+650[12] of TeV flares that cannot be correlated to X-ray flares have is challenging for both leptonic and hadronic models.

In general HBL and LBL/FSRQ exhibit completely analogous variability with respect to the (different) peak frequencies in their SEDS: both vary more above their respective power peaks. It is just that this makes LBL highly variable in the optical and GeV gamma rays, whereas HBL are highly variable in X rays and in TeV gamma rays (and relatively quiescent in the optical and GeV ranges). Although variability is one of the most remarkable characteristics of blazars, the present theoretical work focuses on the analysis of time-averaged SEDs in order to study emission mechanisms and largely neglects the variability origin and the development of the plasma that supports the non-thermal particles during a flare. Only recently this aspect have been addressed in a work concerning the internal shock model.[20]

3. Conclusion

Past MW campaigns have allowed us to expanded our knowledge of blazars but outstanding still unanswered questions exist about the structure and origin of the blazar jet (e.g. how the jet are produced in the region close to accreting black holes, how it is accelerated and what is its matter content, etc). Other important questions are related to the jet emission mechanism (e.g electrons acceleration mechanism, variability etc.).

We are now starting again a new season favorable to blazar studies both by satellites (e.g. *Swift*, Suzaku, and AGILE and GLAST) and by ground based facilities, and in next years blazars will be observed in all bands almost without energy gaps.

The *Gamma-ray Large Area Space Telescope* (GLAST), scheduled to launch in October 2007, is expected to detect a large numbers of blazars and its Large Area Telescope (LAT) will greatly contribute to study their spectral and time variability.

The LAT instrument is designed to detect γ-ray photons in from 30 MeV to above 300 GeV. It will observe about 20% of the sky at any instant with good sensitivity (see fig. 2) the entire sky will be covered in a few orbits with uniform exposure. This will make possible to investigate blazar variability on timescales 1 day and to perform detailed spectral variation analysis and intrabands delays studies.[4]

To study both spectral shape and variability of blazar in future MW campaigns a regular broadband monitoring of a source is required over time scales longer than the typical 10 days of current multiwavelength campaigns. This can be obtained only if a good coordination between ground and space based observatories will be established. In this way multiepoch SEDs (from radio to TeV) of many sources might be obtained over a long period of time and then we will be able to find correlations among the blazar emission at different energy bands and at different time scales helping us to better understanding the blazar phenomenon.

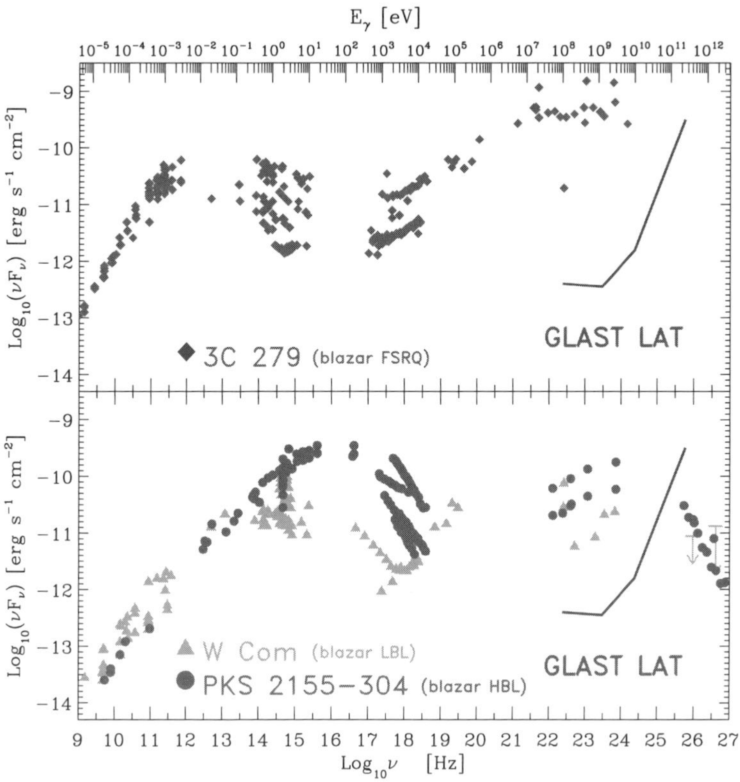

Fig. 2. SEDs of three well studied blazars. The continuous line indicates the GLAST-LAT sensitivity.

It is a pleasure to thank S. Ciprini, L. Foschini, L. Fuhrmann, P. Giommi, E.Massaro, E. Pian, C. Raiteri, G. Tagliaferri, A. Tramacere, M.Villata for continuous interactions and collaborations.

References

1. Aharonian, F.A. et al., 1997. A&A 327, L5.
2. Błażejowski et al.,2005, 630,130
3. Böttcher M., 2006, astro-ph/0608713
4. Carson J.,2006,astro-ph/0610960
5. Fossati G. et al. 1998, MNRAS, 299, 433
6. Hartman, R.C., Bertsch, D.L., Bloom, S.D., 1999. ApJS, **123**, 79.
7. Hartman, R. C., et al., 2001a,ApJ, 553, 683
8. Hartman, R. C., et al., 2001b,ApJ, 558, 583

9. Jorstad, S. G., et al. 2001, ApJ, 556, 738
10. Kellermann K. I., et al. 2004, ApJ, 609, 539
11. Krawczynski H. 2004, NewA, 48, 367
12. Krawczynski H., et al. 2004. ApJ, 601, 151
13. Lister L. M., Homan D. C., 2005, AJ, 130,1389
14. Madejski G. M.,et al. 1999, 521, 145
15. Marscher A. P., & Gear, W. K. 1985, ApJ, 298, 114
16. Padovani P., Giommi P., 1995, MNRAS. 227, 147
17. Padovani P., 2006,astro-ph/0610545
18. Piner B. G., & Edwards P. G. 2004, ApJ, 600, 115
19. Sambruna R. M., Maraschi L., Tavecchio F., et al., 2002, ApJ, 571, 206
20. Spada et al., 2001 MNRAS 325, 1559
21. Tagliaferri G., Ghisellini G.,Ravasio M.,2002, astro-ph/0207017
22. Ulrich M.-H., Maraschi L., & Urry, C.M., 1997, ARA&A, 35, 445
23. Urry, C. M., Padovani P., 1995, PASP, **107**, 803
24. von Montigny C., et al. 1995, ApJ, 440, 525

AGN Observations with the MAGIC Telescope

C. Bigongiari* for the MAGIC Collaboration

Università di Padova and INFN Padova,
Via F.Marzolo 8 - I35131 Padova, Italy
**E-mail: ciro.bigongiari@pd.infn.it*
wwwmagic.mppmu.mpg.de

MAGIC is presently the imaging atmospheric Cherenkov telescope with the largest reflecting surface and the lowest energy threshold. MAGIC concluded its first year of regular observation in April 2006. During this period and the preceding commissioning phase, 25 Active Galactic Nuclei have been observed and VHE γ-ray emission has been confirmed by 4 of them. Two more AGNs have been detected as γ-ray sources with high statistical significance for the first time. We report in this paper the results obtained analyzing data of the detected sources. Temporal and spectral properties of detected signals are shown and discussed.

Keywords: AGN, Blazars, TeV γ-ray astrophysics, Cherenkov Telescope

1. Introduction

MAGIC (Major Atmospheric Gamma Imaging Cherenkov) telescope,[1,2] located on the Canary Island La Palma (2200 m a.s.l., 28.4′N, 17.54′W), is currently the largest imaging air Cherenkov telescope in operation. The MAGIC construction was completed in Fall 2003 and after a commissioning phase of about one year MAGIC started its first regular observation cycle in April 2005. According to detailed simulation of atmospheric showers and detector response the trigger threshold is around 60 GeV for low zenith angle observations[3] while the analysis threshold is about $E_{Th} = 100$ GeV in the same conditions. MAGIC integral flux sensitivity has been calculated from Monte Carlo simulation and results in about 5% of Φ_{Crab} at $E > 100$ GeV and 2% of Φ_{Crab} at $E > 1$ TeV.[3] The angular resolution has been estimated applying the DISP method to Crab data and results in about 0.1° for γ-ray events with $E > 200$ GeV.[4] The accuracy in the determination of the point-source position improves as the square root of the number of collected events and is ultimately limited by tracking accuracy

($\simeq 0.02°$).[5] The energy resolution has been estimated from Monte Carlo data and results in $\Delta E/E \simeq 30\%$ at $E = 100$ GeV and $\Delta E/E \simeq 20\%$ for $E > 1$ TeV.[6] The systematic errors on the measured flux were estimated to be around 50% for the absolute flux level and 0.2 for the spectral index.

2. Observed Blazars

MAGIC, during its first year of regular data taking, observed a sample of Blazars, mainly HBLs, at redshifts $z < 0.3$. This sample was chosen selecting northern Blazars with the highest expected VHE (here defined as $E > 100$GeV) fluxes according to leptonic and hadronic models of γ-ray emission. Moreover, considering possible correlations between VHE emission and optical/X-ray emission, MAGIC performed Target of Opportunity observations whenever alerted by optical/X-ray telescopes. We present here the spectral properties and the temporal behavior of the detected sources: Mkn 421, Mkn 501, Mkn 180, 1ES1959+650, 1ES1218+304 and PG1553+113.

2.1. *Markarian 421*

Mkn 421 (redshift $z = 0.030$) was the first extragalactic VHE γ-ray source detected by Whipple[7] in 1992. Many observations of this source were performed since then showing flux variations larger than one order of magnitude and flares with doubling times as short as 15 minutes.[8] A significant correlation between X-ray and γ-ray flux has been detected during multi wavelength campaigns involving Cherenkov telescopes and X-ray detectors.[10]

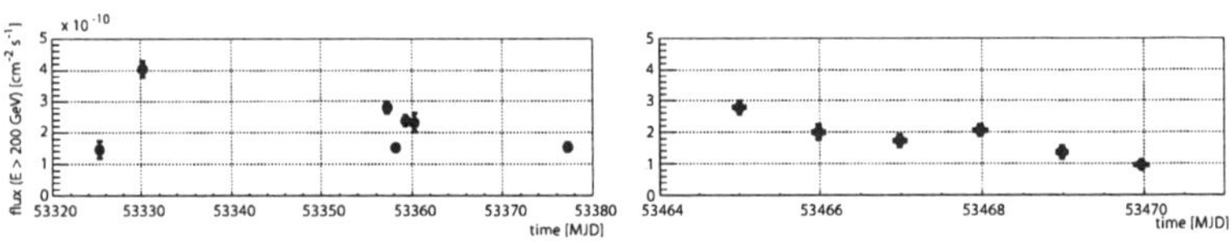

Fig. 1. Light curve for Mkn 421 from November 2004 to April 2005. Each data point is the nightly averaged integral flux above 200 GeV. Left panel: data from November 2004 to January 2005. Right panel: data for April 2005.

MAGIC observed Mkn 421 for 19 nights and an overall observation time of 15.5 hours. All the data were taken at zenith angle below (30°) with the only exception of 1.5 hours taken in December 2005 at $42° < ZA < 55°$ during simultaneous observations with H.E.S.S..[11]

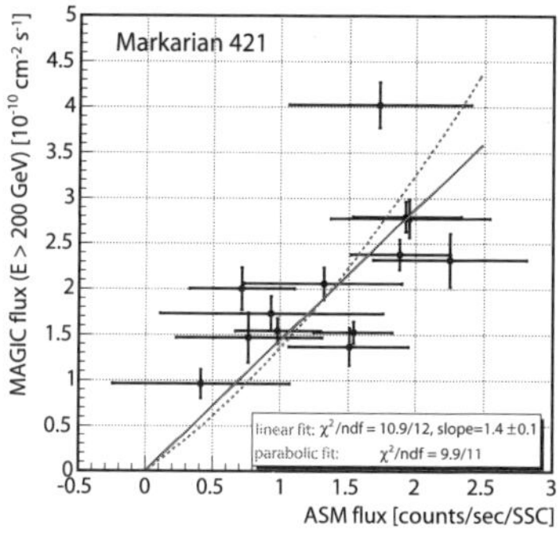

Fig. 2. Correlation plot between VHE γ-ray flux above 200 GeV measured by MAGIC and and X-ray counts by ASM-RXTE for Mkn 421.

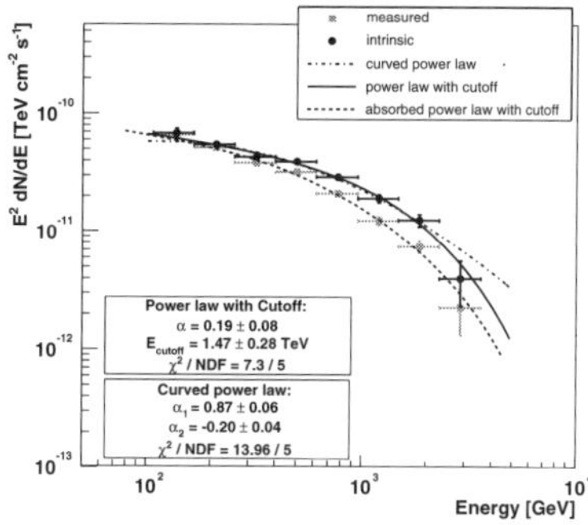

Fig. 3. Measured (gray squares) and de-absorbed (black circles) energy flux of Mkn 421.

During MAGIC observations Mkn 421 flux above 200 GeV ranged from 0.5 to 2 Crab units (see Fig. 1). Significant flux variations up to a factor four overall and up to a factor two between successive nights can be seen. A clear correlation between the X-ray flux, measured by the All-Sky-Monitor on-board the RXTE satellite, and the VHE γ-ray flux measured by MAGIC can be seen in Fig. 2. The energy density distribution of gammas from Mkn 421, that is the differential photon spectra multiplied by E^2, is shown in Fig. 3 both for the measured spectrum and the de-absorbed one (i.e. corrected for the effect of extragalactic absorption). The de-absorbed spectrum is curved, clearly indicating that the curvature in the measured spectrum is not caused by the absorption of the VHE γ-rays by the EBL photons but has an intrinsic origin. For further details about Mkn 421 data analysis see.[12]

2.2. *Markarian 501*

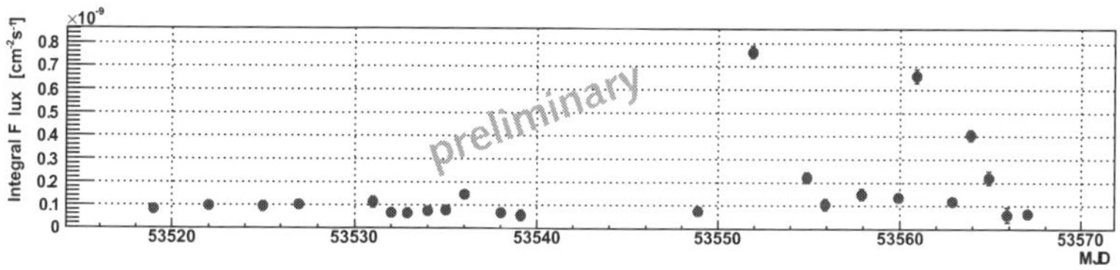

Fig. 4. Light curve for Mkn 501 from June to July 2005. Each data point is the night averaged integral flux above 200 GeV.

Mkn 501 (redshift $z = 0.034$) was the second extragalactic VHE γ-ray source discovered by Whipple[13] in 1996. Following observations in 1997 showed that Mkn 501 integral flux above 1 TeV can reach 10 Crab units[14] and can drastically change on timescales of 0.5 days.

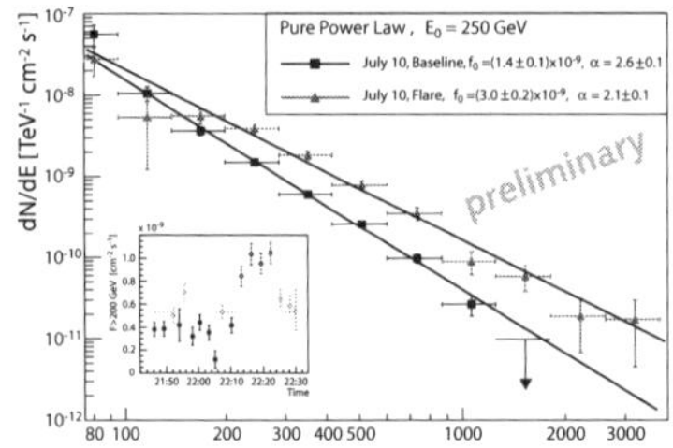

Fig. 5. Differential energy distribution of Mkn 501 events recorded on 10th July 2005. The rapid variation of the flux level and corresponding change in the shape of the energy spectrum is clearly visible.

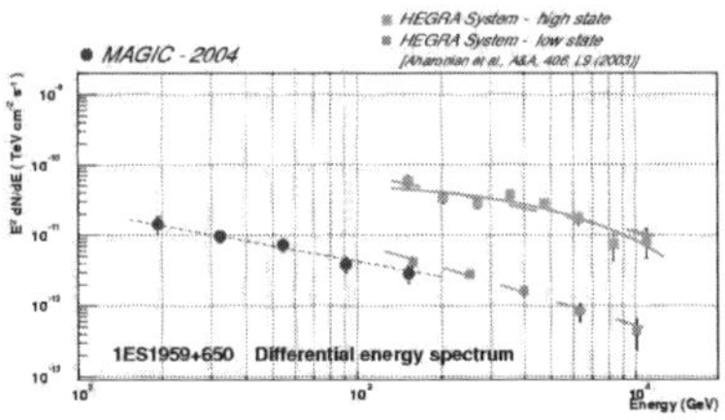

Fig. 6. Differential energy spectrum of 1ES1959+650. For comparison, the differential energy spectra measured by HEGRA when the source was in two different activity levels are also shown.

MAGIC observed Mkn 501 between June and July 2005 for a total of 24 nights and 32.2 hours of observation time. Mkn 501 was in a rather low flux state (integral flux above 200 GeV around 0.4 Crab units) when MAGIC started its observation. Suddenly, on 30th Jun, its flux reached 4 Crab units, see Fig 4. This flare stimulated further observations which were performed in the following days also in the presence of moonshine to extend the time coverage. The source was found in high state (integral flux above 2 Crab units) on two more nights. In particular a flare with doubling time as short as 5 minutes or less was detected on the night of 10th July 2005, see the inlay in Fig. 5. The high source flux together with the MAGIC high sensitivity allowed the measurement of the source spectrum in time intervals as short as 10 minutes. A significant hardening of the spectrum as the flux grows was found, confirming previous indications by Whipple.[13] It is worth noticing that this is the first time that spectral hardening has been measured on time scales of $\sim$ 10 minutes. A detailed publication on the results of the Mkn 501 data analysis is in preparation.

2.3. *Markarian 180*

Mkn 180 (1ES1133+704) at redshift $z = 0.045$ is the second extragalactic γ-ray source discovered by MAGIC. Previously this source had been observed by HEGRA and Whipple collaborations but these observations resulted only in upper limits on the VHE γ-ray flux.[15,16] MAGIC observed Mkn 180 for 9 nights after an alert received by the KVA telescope, on 23rd March 23 2006. In the following nights MAGIC observed this source for 14.4 hours. Only 11.1 hours survived to the quality cuts applied to remove runs with unusual trigger rate, usually related to not optimal atmospheric conditions. This sample was enough to find a clear signal at 6.5σ level with 271 excess events and to measure the integral flux above 200 GeV, $\Phi(E > 200\text{GeV}) = (2.3 \pm 0.7) \times 10^{-7} m^{-2} s^{-1}$, which corresponds to 10% of the Crab Nebula flux as measured by MAGIC. No evidence of flux variation was found on a daily scale. The differential energy distribution, shown in figure 7 with filled

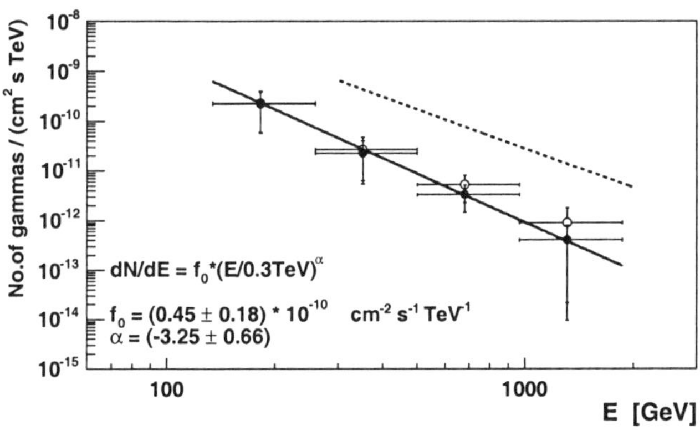

Fig. 7. Differential energy spectrum of Mkn 180, measured by MAGIC (full circles) and corrected for the effect of extragalactic absorption (open circles). The black line represents a power-law fit to the measured spectrum. The fit paremeters are listed in the figure. For comparison, the Crab Nebula spectrum measured by MAGIC is also shown (dotted line).

points, can be fitted by a power law with spectral index $\alpha = 4.2 \pm 0.7$. In the same figure also the de-absorbed spectrum (i.e. corrected for the effect of the extragalactic absorption) is shown (open circles). For further details on this analysis see.[17]

2.4. *1ES1959+650*

The first hint of VHE γ-ray emission from 1ES1959+650 ($z = 0.047$) was claimed by the Seven Telescope Array collaboration in 1998.[18] This claim was later confirmed both by Whipple and HEGRA collaborations[19,20] which observed this source in May 2002 when its X-ray flux was much higher than the usual level. During observations in 2002 a so-called *orphan flare*, i.e. a VHE γ-ray activity in the absence of high activity in X-rays, was observed. Orphan flares are very interesting because they could be an indication of hadronic acceleration in Blazars. They are not expected, in fact, within the SSC model. MAGIC observed 1ES1959+650 when it was still in the commissioning phase, therefore not at its best performances. Nevertheless a clear signal at 8.2σ significance level was found and the energy spectrum was measured down to 180 GeV for the first time. The energy spectrum between 180 GeV and 2 TeV can be well fitted with a power law with a photon index $\alpha = 2.72 \pm 0.14$ and is consistent with the one measured by HEGRA,[20] see Fig 6. The integral VHE γ-ray flux above 180 GeV resulted in $(3.73 \pm 0.41) \times 10^{-7} m^{-2} s^{-1}$, in agreement with the low state flux measured by HEGRA. For further details of the analysis of these data and its results see.[21]

2.5. *1ES1218+304*

The Blazar 1ES1218+304 at redshift $z = 0.182$ is the first extragalactic VHE γ-ray source discovered by MAGIC. This source has been observed by Whipple since the discovery of Mkn 501 but these observations leaded only to an upper limit on the VHE flux.[22] HEGRA collaboration also observed this source, but again no detection was claimed and only an upper limit on the γ-ray flux was published.[23]

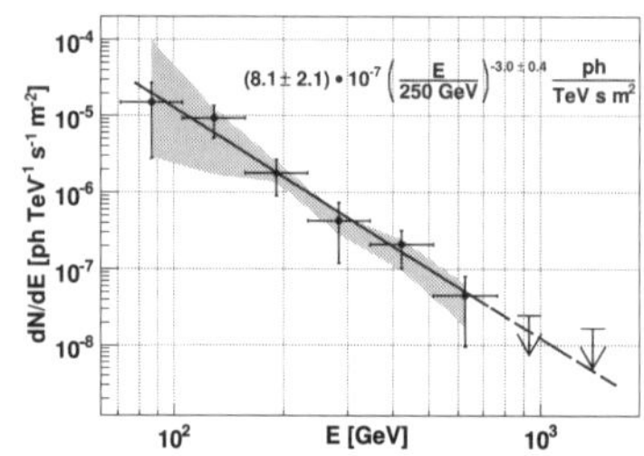

Fig. 8. Differential energy spectrum of 1ES1218+304. The upper limits correspond to 90% confidence level. The grey-shaded region shows the systematic error due to initial MC spectrum analysis cuts.

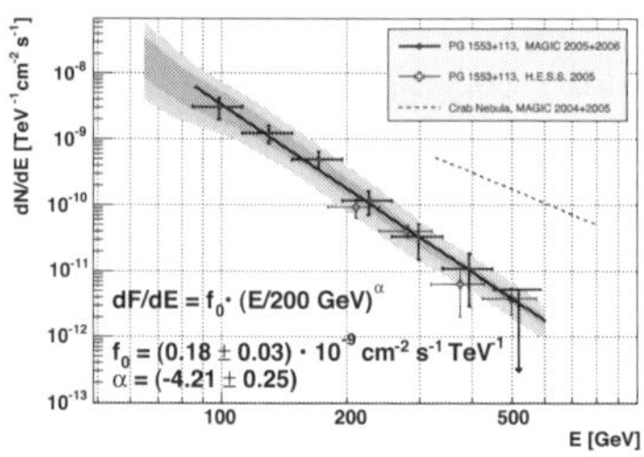

Fig. 9. Differential energy spectrum of PG1553+113 as derived from the combined 2005 and 2006 data. The grey-shaded region shows the systematic error due to initial MC spectrum and analysis cuts.

MAGIC observed 1ES1218+304 for 8.2 hours during seven nights in January 2005. An excess of 560 events with statistical significance of 6.4σ was found. The night-by-night γ-ray light curve didn't show any statistically significant variations. The energy spectrum between 80 GeV and 600 GeV can be well fitted by a power law with photon index $\alpha = 3.0 \pm 0.4$. The integral flux above 100 GeV is $\Phi(E > 100\text{GeV}) = (8.7 \pm 1.4) \times 10^7 m^{-2} s^{-1}$ and is below the upper limits at higher energies determined in the past. For details about these data analysis see.[24]

2.6. *PG1553+113*

The first evidence of VHE γ-ray emission from PG1553+113 (redshift unknown, $z > 0.09$) was claimed in 2005 by H.E.S.S. collaboration showing an excess at 4σ level.[26] MAGIC started to observe this source in 2005 and, motivated by a strong hint of signal in the 2005 data, continued in 2006 for an overall observation time of 18.8 hours. A very clear signal was detected with a significance of 8.8 σ. There is no evidence of γ-rays short term variability, but a factor of three change in the flux level from 2005 to 2006 was found. The combined 2005 and 2006 differential energy spectrum for PG1553+113 is well described by a pure power law with a photon index $\alpha = 4.2 \pm 0.4$, in good agreement with H.E.S.S. result in the overlapping energy range. For details of the analysis and results see.[27]

3. Conclusions and Outlook

An overview of the extragalactic VHE γ-ray sources detected by MAGIC until May 2006 is given. In this period MAGIC detected VHE γ-ray emission by 6 extragalactic sources: Mkn 421, Mkn 501, Mkn 180, 1ES1959+650, 1ES1218+304 and PG1553+113. Two of them, 1ES1218+304 and Mkn 180, have been discovered by MAGIC while PG 1553+113 has been confirmed as γ-ray source at high significance level after the first hint of signal by H.E.S.S. The energy spectrum of Mkn 421 has been measured down to 100 GeV for the first time showing a de-absorbed spectrum which clearly flattens toward 100 GeV. Mkn 501 has been caught in flaring states in 3 nights with flux higher than 2 Crab units. Flux doubling time down to 5 minutes has been observed and spectral hardening as the flux increases has been measured on 10 minutes time scale. Gamma-ray emission from Mkn 180 has been discovered during an optical outburst indicating a possible correlation between γ-ray and optical emission. The energy spectrum of 1ES1959+650 has been measured down to 180 GeV, well below previous measurements by HEGRA, when the source was in a low activity state.

Re-observations of the presented sources as well as the analyses of further observed Blazars are ongoing.

Collaborations with observatories in other spectral ranges have proved to be very fruitful and will be extended and strengthened in the near future.

4. Acknowledgments

We would like to thank the IAC for the excellent working conditions at the ORM in La Palma. The support of the German BMBF and MPG, the Italian INFN, the Spanish CICYT, the ETH research grant TH 34/04 3, and the Polish MNiI grant 1P03D01028 is gratefully acknowledged.

References

1. Baixeras, C., et al., Nucl. Instrum. Meth. Phys. Res. A, **518**, 188 (2004)
2. Cortina, J. et al., In Proc. 29th Int. Cosmic Ray Conf. (Pune), **5**, 359 (2005)
3. P. Majumdar et al., In Proc. 29th Int. Cosmic Ray Conf. (Pune), **5**, 359 (2005)
4. E. Domingo Santamaria et al., In Proc. 29th Int. Cosmic Ray Conf. (Pune), **5**, 359 (2005)
5. B. Riegel et al., In Proc. 29th Int. Cosmic Ray Conf. (Pune), **5**, 359 (2005)
6. Wagner, R.M. et al., In Proc. 29th Int. Cosmic Ray Conf. (Pune), **4**, 163 (2005)
7. Punch, M. et al., Nature **358**, 477-478 (1992)
8. Gaidos, J. A. et al., Nature **383**, 319-320 (1996)
9. Aharonian, F. et al., A&A **437**, 95-99 (2005)
10. Krawczynski, H. et al., ApJ **559**, 187-195 (2001)
11. Mazin, D. et al., In Proc. 29th Int. Cosmic Ray Conf. (Pune), **4**, 331 (2005)
12. Albert, J. et al., Subm. to ApJ, astro-ph/0603478 (2006)
13. Quinn, J. et al., ApJ **456**, L83 (1996)
14. Aharonian, F. et al., A&A, **342**, 69-86 (1999)
15. Aharonian, F. et al., A&A **421**, 529-537 (2004)
16. Horan, D. et al., ApJ **603**, 51-61 (2004)
17. Albert, J. et al., ApJ in press, astro-ph/0606630 (2006).
18. Nishiyama, T. et al., In Proc. 26th Int. Cosmic Ray Conf. (Salt Lake City) **3**, 370 (1999)
19. Holder, J. et al. ApJ **583**, L9-L12 (2003)
20. Aharonian, F. et al., A&A **406**, L9 (2003)
21. Albert, J. et al., ApJ **639**, 761-765 (2006)
22. Horan, D. et al., ApJ **603**, 51-61 (2004)
23. Aharonian, F. , In Proc. 27th Int. Cosmic Ray Conf. (Hamburg), astro-ph/0112314 (2001)
24. Albert, J. et al., ApJ **642**, L119-L122 (2006)
25. Aharonian, F. et al., Nature **440**,1018 (2006)
26. Aharonian, F. et al., A&A **448**, L19-L23 (2006)
27. Albert, J. et al., Subm. to ApJ, astro-ph/0606161 (2006)

GAMMA–RAY BURSTS

L. AMATI*

INAF - IASF Bologna,
via P. Gobetti, 101 I-40129 Bologna, Italy
** E-mail: amati@iasfbo.inaf.it*
www.iasfbo.inaf.it

I present a short review of Gamma–Ray Bursts studies, including observations, standard pictures, recent discoveries and the future perspectives of this very intriguing field of modern astrophysics.

Keywords: Gamma–rays: bursts

1. Introduction

The study of Gamma–Ray Bursts is one of the hottest topics in modern astrophysics. Indeed, despite their study started more than 30 years ago and the enormous observational and theoretical progress occurred in the last 10 years, many aspects of these mysterious phenomena and of the sources producing them (the "progenitors") are still obscure. In the following I give a brief review of the observations, interpretations, recent results and future perspective of GRB science. For reasons of space, I do not include figures and a give only a few references; plenty of both can be found in several exhaustive reviews.[1–4]

2. The GRB phenomenon

2.1. *Early observations and BATSE discoveries*

Gamma–Ray Bursts (GRBs) are sudden and unpredictable bursts of hard X / soft gamma–rays with huge intensity, typical durations of tens of seconds and coming from random directions in the sky. They were discovered at the end of the '60s by military satellites and the first scientific paper on this subject was published on an astronomical journal (ApJ) in 1973. During the '70s and '80s, several GRB experiments were carried on-board various satellites, but with poor improvements in the understanding of these phenomena. A major contribution came in the '90s from the NASA BATSE experiment (25-2000 keV) on-board CGRO (1991-2000). This instrument was based based on NaI scintillator detector units covering a 4π FOV, and gave the first good characterization of the GRB phenomenon: most of the flux is detected from 10-20 keV up to 1-2 MeV; the measured rate by an all-sky experiment on a low Earth orbit satellite like CGRO is 0.8 / day (corresponding to an estimated true rate $\sim$2 / day); the fluences (i.e. the average flux multiplied by the duration) range between $\sim 10^{-7}$ and $\sim 10^{-4}$ erg/cm2; diverse and unclassifiable light curves and a bimodal distribution of durations (short, from a few ms to 1–2s, and long form 1-2 s to $\sim$1000s); short GRBs tend to be spectrally harder than long GRBs; non thermal spectra typically described by a smoothly broken power-law which can be modeled by the so-called "Band model" (parameters: α = low-energy index, β = high-energy index, E_0 = break energy, $E_p = E_0 \times [2 + \alpha]$ = peak energy of the $\nu F \nu$ spectrum); typical hard to soft spectral evolution throughout the whole GRB or hard to soft during each pulse; isotropic distribution of GRBs directions and paucity of weak events with respect to homogeneous distribution in Euclidean space (hints to cosmological origin of GRBs). Moreover, CGRO/EGRET detected VHE (from 30 MeV up to 18 GeV) photons for a few GRBs, showing that VHE emission can last up to thousands of s after GRB onset. VHE average spectrum of 4 events detected by EGRET is well described by a simple power-law with index 2 , consistent with extension of low energy spectra. GRB 941017, as measured by EGRET-TASC, shows a high energy component inconsistent with low–energy emission.

2.2. *BeppoSAX: the afterglow era*

A "revolution" in the GRB field occurred in 1997 thanks to the discoveries of the Italian–Dutch satellite BeppoSAX (1996-2002). The payload of this mission was composed by a set of narrow field instruments (NFI), which included X-ray focusing telescopes operating in the 0.1-10 keV energy band, and a set of wide field instruments, the Wide Field Cameras (WFC, 2 units, proportional counters + coded mask, FOV 20x20 deg., 2-28 keV energy band) and the GRB Monitor (GRBM, 4 units, CsI scintillators, large FOV, GRB triggering, 40-700 keV energy band). The co–alignment of 2 units of the GRBM with the WFC made it possible the simultaneous detection of GRBs in hard and soft X–rays, thus providing GRB localization with unprecedented accuracy (few arcmin). The fast (few hours) follow up of the WFC error boxes with the NFI led to the discovery of GRB X-ray afterglow, i.e. fading emission following the "prompt" GRB emission. The X–ray afterglow measurements performed first by by BeppoSAX and, later, by Chandra and XMM, put in evidence a typical power-law decay (even if in some cases small bumps were detected) and power-law spectra. The very accurate localization of the X–ray afterglow sources (1 arcmin or less) allowed also the follow–up with optical and radio telescopes, leading to the discovery of the first optical, IR, radio GRB counterparts. This low energy afterglow emission shows a temporal behaviour similar to that observed in X–rays, even if "bumps" and "breaks" have been observed for several events, as will be discussed in next sections. All this new phenomenology allowed to improve our understanding of GRBs, but the most striking breakthrough was the first measurements of GRB redshifts through spectroscopy of optical counterparts and/or host galaxies. This solved the long–standing question concerning GRB distance and energetics. Up to now there are ~80 long GRBs (and a few short) with measured redshift and all of them (except the peculiar GRB 980425) lie at cosmological distances (z = 0.0331–6.3). The radiated energy is huge, up to $\sim 10^{54}$ erg if assuming isotropic emission.

3. The standard picture and the basic physics

The observational evidences summarized above are at the basis of the standard picture for GRB sources. As was already argued in the 70s, the ms time variability plus the huge energy budget plus the detection of GeV photons point to a plasma occurring ultra–relativistic ($\Gamma > 100$) expansion. This kind of source is commonly called "fireball". The non thermal spectra suggest synchrotron and/or Inverse Compton as dominant emission mechanisms. In the standard scenario (synchrotron shock model, SSM) the GRB emission originates in shocks between shells of the fireball moving at different velocities (internal shocks). In these collisionless shocks, electrons are accelerated to a power–law distribution (e.g. by Fermi acceleration), a very intense magnetic field is generated and electrons radiate by synchrotron. This synchrotron radiation maybe the direct responsible of the X and gamma–ray radiation, or may produce lower energy photons that are than up–scattered to high energies by Inverse Compton (IC) interaction with the relativistic electrons. Also, thermal emission, produced when the fireball becomes optically thick, may contribute to the early phases of the GRB or give rise to a "precursor", i.e. a small burst occurring before the main event. GRB synthesis models show that internal shocks can reproduce the complex GRB light curves and their variability. The afterglow emission is supposed to be generated by the interaction (deceleration) of the fireball with the ISM (external shock). Synchrotron is thought to be the dominant emission mechanism, even though an IC contribution may be present. The evidence that the width of GRB pulses increases as the energy band decreases is consistent with synchrotron shock models predictions; also, many BATSE (25-2000 keV) and BeppoSAX(2-700 keV) spectra are consistent with the predictions of synchrotron shock emission models (SSM). However, spectral evolution analyses show that for a fraction of GRBs the low energy photon index measured during the first phases of the emission is inconsistent with optically thin synchrotron emission (i.e. $\alpha > -0.67$). This evidence may be explained by invoking different emission mechanisms: quasi–saturated comptonization, Compton drag, synchrotron emission with small pitch angle, syncrotron self-absorption, significant contribution of thermal emission from the photosphere of the fireball. An intriguing evidence is that short and long GRBs show different distributions of spectral parameters: short GRBs tend to be spectrally harder. In particular, it has been observed that the spectra of the first 2 s of long GRBs resemble those of short GRBs. This could indicate that the spectrally hard early phase of the GRB emission is mainly produced by a different emission mechanism with respect to the later longer and softer emission.

The power-law decay and power-law spectrum typical of afterglow emission is consistent with the predictions of synchrotron radiation produced in external shock. Also, in some cases multiwavelength spectral analysis from radio to X–rays (SED) show consistency with synchrotron predictions, as found, e.g., for GRB 970508 . However, there are also cases for which the SED at a given time cannot be described by synchrotron emission alone, as in the case of GRB 000926 and GRB 010222, in which there is evidence of a significant contribution of Inverse Compton to the X–ray agterglow emission.

4. Unvealing the progenitors

The standard picture for GRB emission, schematically summarized above, assumes that the ultra–relativistic fireball is originated by the release of a huge amount of energy (up to $\sim 10^{54}$ erg, or more, for bright long GRBs) in a compact region of space (radius of $\sim 10^5$ cm) lasting several tens or hundred of seconds (to explain the durations of long GRBs). The "engine" producing this release of energy is commonly thought to be formed by a stellar mass black hole fastly accreting a torus of matter; different mechanisms have been proposed for the extraction of the gravitational energy of this system and its conversion into a plasma (pairs plus a small fraction of baryons) expanding at relativistic velocities. The most common hypothesis on the nature of the progenitors of both long and short GRBs all predict the formation of such systems. The critical points are the capability of producing such a huge energy release, the duration of the engine activity, the mechanisms and efficiency of conversion of gravitational energy into kinetic energy the fireball, the efficiency of conversion of fireball kinetic energy into radiation.

4.1. *Circum–burst environment*

The properties of the circum–burst environment can give us hints on the nature of the progenitors of both long and short GRBs. The discovery, by BeppoSAX, of a transient absorption feature in the prompt X–ray spectrum of GRB 990705, interpreted as a redshifted neutral iron K–edge or resonant scattering of Fe XXVI is an evidence of a very iron-enriched absorbing medium. This hypothesis is further supported by the detection (even if still debated) of transient emission lines in X–ray afterglow spectra, interpreted as redshifted Fe fluorescence / recombination lines. The existence of dark GRBs (showing X–ray afterglow emission but no optical counterpart) may be interpreted as due to high oscuration by dust surrounding the GRB site or as inefficient external shock due to low density medium. Also, afterglow decay spectral slopes, their evolution and their relations (closure relations) depend on circum-burst environment properties (e.g., homogeneous, wind). In several cases it has been found that an emission model based on external shock into a medium with a wind profile (as expected if the progenitor is a supernova) fits better the data. Finally, deviations of afterglow light curve from monotonic power-law decay, observed in some cases, may be originated by refreshed shocks due to inhomogenities of the circum-burst environment.

4.2. *The GRB – SN connection*

The hypothesis that GRB may be originated by peculiar kinds of supernovae (SN) was proposed already in the '70s and reinforced when the first redshift estimates put in evidence the huge energy released by these events. But the firts direct evidence of a GRB–SN connection came only in 1998 with GRB 980425 / SN1998bw. GRB 980425 was a normal GRB detected and localized by BeppoSAX WFC and NFI, but it occurred in temporal/spatial coincidence with a type Ib/c ("core collapse") SN at z = 0.008 (chance prob. 0.0001). Further evidences of a GRB/SN connection came from bumps in the optical afterglow light curves of some GRBs, showing spectra resembling that of SN1998bw (the most famous case is GRB 030329). The few SN associated with GRBs based on spectroscopic evidence are of Type Ib/c, are very bright ($M_V \sim -19 \div -19.5$) and belong to the "hypernova" class, i.e., they are characterized by ejecta velocities about 10 times higher than those of "normal" type Ib/c SNe. Also the typical characteristics of the host galaxies(blue, usually regular and high star forming) of long GRBs and the location of these events inside star forming regions are suggestive of an association between GRBs and core collapse SN population.

4.3. *Short GRBs*

Only very recently (2005) it has been possible to accurately localize and follow–up a few short GRBs and thus to discover their afterglow emission and optical counterparts and to estimate their redshifts and energetics. While their afterglow emission shows properties similar to those of long GRBs, short GRBs seem to be located at lower redshifts (<1) and to be less energetic ($<\sim 5\times10^{51}$ erg). The host galaxies of short GRBs, contrary to long ones, show different morphologyies (spiral, elliptical, irregular). Moreover, short GRBs are often located in the outskirts of their host galaxy and away from star forming regions. These evidences, combined also with the emission properties summarized in the previous section, further support a different origin of short and long GRBs.

4.4. *Main progenitors models*

The standard model for the progenitors of long GRBs is the so-called "collapsar", i.e. a massive (more than $\sim$30 M_{sun}) and fastly rotating star undergoing an asymmetric collapse along the rotation axis, whose core directly collapses into a black hole surrounded by a torus of matter ($\sim$0.1 M_{sun}). As mentioned above, the very fast accretion of the torus by the black–hole can release a huge amount of energy through extraction of gravitational energy or neutrino emission. This scenario can explain several of the evidences for long GRBs summarized above: energy budget up to $>10^{54}$ erg, long duration, metal rich (Fe, Ni, Co) wind circum-burst environment, location in star forming regions, association with core–collapse SNe. In addition, in several cases breaks in the optical afterglow light curves were discovered. This fact has been interpreted as an evidence of collimated emission, even though other explanations, like, e.g., transition from relativistic to non–relativistic expansion, have been proposed. The jet angles derived from the optical break time, are of the order of few degrees and the collimation–corrected radiated energies cluster around $\sim 5\times10^{51}$ erg. These values are in agreement with the predictions of collapsar models. Alternative scenarios include: the supra–nova, in which the core of the star collapses into a vary fastly rotating neutron stars, which collapses into a black hole and produces the GRB weeks or months after the SN explosion; the transition from neutron star to quark star; the cannon–ball model, in which the accretion of the torus by the black hole does not produce a fireball but blobs of plasma, which emit by IC interaction with the UV photons emitted by the SN; the EMBH model, in which the core of the star collapses into an electrically charged black hole, leading to the formation of a dyadosphere, which gives rise to the relativistic, and spherical, fireball; the precessing jet scenario, in which the observed GRB phenomenology is not due to an explosive and fastly transient progenitor but to a source emitting a very highly collimated jet which is precessing and thus moving inside and outside our field of view.

The most popular scenario for the progenitors of short GRBs is the "merger", i.e. the production of a black–hole, and a torus of matter being accreted by it, due to the coalescence of two compact objects (neutron star – neutron star, neutron star – black–hole, neutron star – white dwarf). Simulations show that this kind of progenitor can explain the energy budget up to 10^{51} - 10^{52} erg) and the durations (<2 s) typical of short GRBs. In addition, these systems are expected to be located far away from their formation site, mainly because of the kick–off received during the SN explosions that originated the compact objects, in agreement with their locations away form star forming regions and in the outer regions of the host galaxy. Also, a clean uniform circum-burst environment is expected. Recently, the association of, at least some, short GRBs with giant outbursts from Soft Gamma Repeaters has been proposed, but up to now there is no convincing evidence supporting this hypothesis.

5. Adding pieces to the puzzle

5.1. *Further and recent observational evidences*

One of the most intriguing recent discoveries is the existence of a sub–class of GRBs called X–Ray Flashes (XRF): GRBs with only (or predominant) X-ray emission and very low peak energy E_p. These events were discovered by BeppoSAX at the end of the '90s and studied deeply during the first half of this decade, with HETE–2, a mission dedicated to GRBs carrying on–board a GRB detector (based on NaI scintillators), an X–ray camera (gas proportional counters + 1-D coded mask) and a soft X–ray camera (CCD + 1-D coded mask), thus allowing arcmin localizations of GRBs (and fast dissemination of them) and characterization of the X–ray emission of GRBs. Normal GRBs, X–ray Rich GRBs (XRRs) and XRFs are found to be in the ratio 1:1:1; GRBs, XRRs and XRFs form a continuum in the E_p–fluence plane, an evidence of a common origin. Recent redshift estimates for these events are in the range 0.1–1, thus showing that their softness is likely not due to high redshift. The most popular explanations include: fireball with high baryon loading, inefficient internal shocks due to low contrast of Γ between colliding shells with respect to fireball bulk Γ, off–axis viewing effects.

Another relevant issue is that of optical flashes simultaneous to the prompt X–ray / gamma emission, detected in a few cases thanks to robotic telescopes. The first case was GRB 990123, a BeppoSAX event for which ROTSE detected optical emission simultaneous to the GRB. The light curve of this optical flash did not follow the high energy light curve and showed a monotonic decay, suggesting a different origin for the two emissions. The most popular interpretation is that it was originated by the "reverse shock", i.e. a shock propagating backward in the rest frame of the expanding shell. In other cases, like, e.g., GRB 041219 (RAPTOR), the optical light curve was found to follow the high energy light curve, suggesting an internal shocks origin. For a recent event, GRB 050401 the light curve of the optical flash (ROTSE-III) does not follow the high energy light curve (as in the case of GRB 990123) and is consistent with the back–extrapolation of the later optical afterglow light curve. Thus, this is likely a detection of the rise and early phase of the optical afterglow.

One of the most debated recent evidences in the field of GRB is the strong correlation between the intrinsic (i.e. in the source cosmological rest frame) peak energy $E_{p,i}$ and the isotropic equivalent radiated energy, E_{iso} . This correlation was found in 2002 based on 12 BeppoSAX events with known z, and it was confirmed and extended to XRFs by subsequent observations. It has important consequences for the physics of prompt emission, GRB/XRF unification models, GRB jet models, off–axis XRF models, and is commonly used as an input or test output for GRB population synthesis models. In addition, the $E_{p,i}$ – E_{iso} correlation can be used to identify, and understand, different sub–classes of GRBs and, combined with other observables, it has been used to build GRB redshift estimators, even if this topic is presently highly debated.

Very intriguing is also the possible existence a class of local sub–

energetic GRBs. The first evidence comes from the peculiar characteristics of GRB 980425, the "proto–type" event of the GRB/SN connection. Indeed, this event is much closer than the other GRBs with known redshift (z=0.008) and shows a very low energetics, both from the point of view of the X/gamma–rays prompt emission (~10^{48} erg) and radio afterglow. Another event, GRB 031203, also associated with a SN, 2003dh and relatively close (z=0.105) shows a very low prompt and afterglow energetics. Both events are inconsistent with the $E_{p,i}$ – E_{iso} correlation (even though for GRB 031203 this is still debated), further suggesting that they belong to a peculiar class of GRBs. The most common explanation of the peculiar properties of these events is that they were normal GRBs seen very off–axis. In this case, their measured very low energetics and deviation from the $E_{p,i}$ – E_{iso} correlation would be due to relativistic Doppler and beaming effects.

Concerning very high energy (VHE) emission, TeV emission from GRB 970417a detected by MILAGRITO was recently (2003) reported. No other detections were made at these energies for other several tens of events detectable by MILAGRITO, MAGIC and MILAGRO.

Finally, there was a claim of a very high degree of polarization (80±20 %) in the prompt emission of GRB 021206 by RHESSI, but this was not confirmed by subsequent analysis. Evidence of linear polarization in BATSE events GRB 930131 (>35%) and GRB 960924 (>50%) has been recently found based on the analysis of Compton scattering of GRB radiation off Earth's atmosphere. The polarization of optical afterglow emission is less controversial and has been detected at a level of ~5–10% for a few GRBs (consistent with the expectations of models based on synchrotron and IC emission mechanisms combined with the geometry of the jet and particular viewing conditions).

5.2. *The Swift era*

Swift is a NASA mission dedicated to GRB studies launched at the end of 2004 and realized and managed by an USA / Italy / UK consortium. The payload includes a GRB detector / localizator (CZT+coded mask, 15–350 keV, wide FOV, arcmin ang. res.), an the X–ray telescope (X-ray optics, 0.3-10 keV, arcsec ang.res.) and an optical–UV telescope (sub-arcsec ang.res. mag 24 in 1000 s). The spacecraft has unprecedented slewing capabilities allowing it to point the X–ray telescope to the few arcmin GRB error boxes computed by the GRB detector within 1-2 minutes at most. Thus, Swift can study the transition between prompt and afterglow emission and the afterglow onset (BeppoSAX, Chandra and XMM could be on target only a few hours after the GRB) and allows very fast and precise follow–up observations with optical and radio telescopes, which result in an increased number of counterparts detection and of redshift estimates.

Concerning the transition from prompt to afterglow emission, Swift has discovered a completely, and mostly unexpected, phenomenology: the typical early afterglow light curve is composed by an initial very steep decay lasting a few tens of seconds at most, a subsequent flat decay (i.e. a power–law decay with a slope much lower than the later afterglow emission) lasting a few thousands of seconds, early break and start of the steeper power–law decay observed by previous satellites. Also, in ~50% of GRBs X–ray flares superimposed to the afterglow light curves were detected. Several interpretations have been proposed for these features, requiring significant changes and improvements to the standard models summarized above and adding new hints to the nature of the progenitors.

The very accurate GRB localizations by Swift and they very fast dissemination to optical telescopes is allowing a substantial increase of optical counterparts detections and thus of redshift estimates. In particular, the redshift distribution of Swift GRBs is different from that of pre–Swift events, with an average of z~2.8 (against z~1.6 found for former events). This puts in evidence that the bulk of GRB population is at higher redshift than thought before and that selection effects play a significant role in the sample of GRBs with known redshift.

Another breakthrough of Swift was the first detection of afterglow emission of short GRBs, leading also to the first redshift estimates for this elusive events (e.g., GRBs 050509b, 050709, 050724, 050813, 051221, 060502b). As already mentioned above, the afterglow emission of short GRBs does not show significant differences with that of long GRBs, including the early afterglow features and flares. Instead, with respect to long events, they are located at lower redshifts (~0.1–1) and with a distribution inconsistent with that predicted for events following the star formation rate evolution (while that of long GRBs is). Other differences with long GRBs include a lower energetics (E_{iso}~10^{49} – 5×10^{51}), a different morphologies of host galaxies and the inconsistency with Ep,i-Eiso correlation. All these evidences further support the hypothesis, summarized above, of a different origin with respect to long GRBs.

Finally, Swift detected and studied a few very peculiar events which are challenging the standard picture for GRBs; the most intriguing are GRB 060218 and GRB 060614 . The first is the closer event (z=0.0331) after GRB 980425 and shares two important properties with it: it shows a prominent association with a SN (2006aj) and it is sub–energetic. However, contrary to GRB 980425, it is consistent with the $E_{p,i}$ – E_{iso} correlation, suggesting that the sub–energetic nature of GRB 0606218 may be true and not due to viewing angle effects, as proposed for GRB 980425 and the other possible sub–energetic event GRB 031203. Thus, GRB 060218 may be the proto–type event of a class of a population of local and faint GRBs, whose occurrence rate could be as high as 1000 times that of normal cosmological GRBs. Moreover, Swift observed a long lasting (thousands of seconds) thermal X–ray emission associated with GRB 060218 , which has been interpreted as the first detection ever of a SN shock break–out. The other event, GRB 060614, is important because it is a long GRB for which the limits to the magnitude of a possible associated SN exclude that it has been originated by a bright hypernova, if not to exclude at all its association with a SN. This suggests that long GRBs may originated by different progenitors and stimulated the proposal of a new scheme for the classification of short and long GRBs, not based on the duration but on the presence or absence of an associated SN.

6. Future perspectives

6.1. *Present and next experiments*

The study of the afterglow emission of GRBs is presently covered by excellent space missions (Swift, XMM, Chandra, HST), which will hopefully continue to operate at least for the nexe few years, and by the largest ground optical (e.g., VLT, GEMINI, NTT) and radio telescopes (e.g., NRAO, VLA). Also, very recent improvements in detection and study of optical prompt emission have been made thanks to the increased number and quality of robotic telescopes (e.g., ROSTE, PROMPT, REM) and frequent prompt and precise localizations by Swift. The situation is presently less favorable for the study of GRBs X/gamma–rays prompt emission, which is very important, e.g., for the understanding of emission mechanisms, study of circum–burst environment, understanding and use of spectral–energy correlations, study of XRFs and sub–energetic GRBs. Swift/BAT has a too narrow energy band useful for spectroscopy, thus major advances in this field are still coming mainly from archival data of HETE-2, BATSE and BeppoSAX. For a few GRBs Konus/Wind spectral data (20-10000 keV) are available, and also Suzaku/WAM and RHESSI are starting to provide spectral information from tens of keV to several MeV. Future GRB experiments are designed to fill, at least partly, this gap and to satisfy the need to combine the Swift capabilities with sensitive prompt emission detectors operating in a broad energy band from 1–2 keV up to 1–2 MeV. A strong contribution, for the high energy emission, is expected from GLAST (GRB monitor from 10 keV to 25 MeV and large UHE large area telescope upt to 300 GeV) and, to a less extent, AGILE. Missions extending the study of GRB prompt emission down to soft X–rays and currently under study include the Chinese-French mission SVOM (formerly ECLAIRs), Spectrum-RG/e-Rosita/Lobster (main contirbutions from Russia, Germany, UK), EDGE (main contributions from Italy, The Netherlands, USA, Japan).

6.2. *GRBs as a tool*

The huge luminosity and their redshift distribution extending up to at least 6.3, make GRBs a powerful tool for cosmology. Of particular interest is their use as standard candles through spectral–energy correlation. Indeed, in the very last years, it has been found that by adding one more observable, like the optical afterglow break time or the high signal time scale, the $E_{p,i} - E_{iso}$ correlation becomes tighter and thus useful for the estimate of cosmological parameters in a way similar to SNe Ia. This can be obtained by comparing the luminosity computed based on the redshift (which depends on the cosmological model assumed) and the luminosity computed from the spectral peak energy through the luminosity – peak energy relation. This method is presently very debated, mainly because it is still based on a low number of events and requires sophisticated statistical approaches to avoid circularity problems.

Another way to exploit GRBs huge intensity and high redshift for cosmology is to use them as background sources for the study of the properties and evolution of the intergalactic medium (IG, WHIM) and of galaxies metallicity by means of X–ray absorption spectroscopy. A mission (EDGE) dedicated to these studies is under study by an international consortium led by Italy and The Netherlands. Finally, the association of long GRBs with SNe, even if with some issues to be clarified (as dicussed above), makes them useful tools to trace the star formation rate up to high redshift and to study the properties of the first generation of stars.

References

1. G. Fishman, *PASP* **107**, 1145 (1995)
2. T. Piran, *Review of Modern Physics* **76**, 1143 (2005)
3. P. Meszaros, *Rep. Prog. Phys.* **69**, 2259 (2006)
4. B. Zhang, *ChJAA* **7**, 1 (2007)

X-ray and GeV flares in GRB light curves

A. Galli

INFN of Trieste, Padriciano 99,
Trieste, 34012, Italy
IASF of Rome-INAF, Via fosso del cavaliere 100,
Roma, 00133, Italy
E-mail: alessandra.galli@ts.infn.it,
alessandra.galli@iasf-roma.inaf.it

L. Piro

IASF of Rome/INAF, Via fosso del cavaliere 100,
Roma, 00133, Italy
E-mail: luigi.piro@iasf-roma.inaf.it

F. Longo

INFN of Trieste, Padriciano 99 ,
Trieste, 34012, Italy
University of Trieste, Via Valerio 2,
Trieste, 34127, Italy
E-mail: francesco.longo@ts.infn.it

N. Omodei

INFN of Pisa, Polo Fibonacci Largo B. Pontecorvo 3,
Pisa, 56127, Italy
E-mail: nicola.omodei@pi.infn.it

G. Barbiellini

University of Trieste, Via Valerio 2,
Trieste, 34127, Italy
E-mail: guido.barbiellini@ts.infn.it

Observations are showing that X-ray flares are very common features in GRB light curves. X-ray flares occur from about 100 s up to thousands of seconds after the burst, when the prompt-to-afterglow transition is taking place. In the context of External Shock, X-ray flares are produced by thick shell fireballs emitted by a long duration central engine activity, and represent the beginning of the afterglow emission. The delayed flare photons are expected to interact with the forward shock electrons by Inverse Compton producing high energy counterparts that potentially will be detected by GLAST LAT. Such high energy components could explain the delayed GeV emission detected by EGRET in GRB 940217. Observations with GLAST LAT will give useful information to constrain the origin of X-ray flares.

1. GRB temporal-spectral evolution: the flare phenomenon

After more than 40 years from Gamma-Ray Burst discovery, the physics of their progenitor is still unclear. A powerful investigation tool is the study of the GRB prompt-to-afterglow transition phase. This phase, occurs from about hundred to thousand of seconds after the burst, and it is characterized by a variety of temporal and spectral behaviours due to the contribution of both prompt and afterglow emission. Most intriguing during this phase is the presence of X-ray flares. X-ray flares are likely to trace the activity of the central engine, thus they can give important information about the physics of the progenitor during the first phases of the burst.

X-ray flares were observed on time scale of hundred of seconds for the first time by BeppoSAX in XRF 011030, XRR 011121 and GRB 011211.[1,2] These flares have spectra softer than that of the preceeding prompt emission and consistent with that of the following afterglow. These flares also connect to the following afterglow with a power law if the origin of the time is shifted to the instant of their appearance. This suggested that X-ray flares can represent the beginning of the afterglow emission.

Before the launch of Swift (November, 2004), due to the few X-ray flare observations it was not possible understand if they were a common feature and if their spectral and temporal behaviuor was representative of a population. Indeed, Swift showed that X-ray flares are a common phenomenon, being present in about one half of its GRB sample.[3] X-ray flares occur also in X-Ray Flashes (e.g. XRF 050406) and in short gamma-ray bursts (e.g. GRB 0507024), and several bursts show multiple flares (e.g. GRB 050607, GRB 060730). Swift confirmed that X-ray flares are globally softer that the prompt γ-ray emission[4] but showed also that they can have a variety of spectral behaviours: in several bursts X-ray flares exhibit hard-to-soft spectral evolution resembling that of the prompt emission,[5] while in other bursts they do not present substantial spectral evolution and have spectrum consistent with that of the afterglow emission (e.g. GRB 050126 and GRB 050219A,[6] GRB 050712[7] and GRB 050904[8]). This suggests that two families of X-ray flares could exist.

2. X-ray flares: Possible scenarios

The important implications related to the origin of X-ray flares have given a big effort to the development of X-ray flares models. We can distinguish between two main groups of models, that respectively, do not require and require long duration engine activity.

Models that belong to the first group are: two components jet,[9] patchy jets,[10] Forward-Reverse shock[11] and external shock with a clumpy medium.[12,13] In these models the predicted flare temporal profile is typically too shallow to explain the fast rise and decay observed in X-ray flares.

The temporal profile of X-ray flares can be explained better by models of the second group. These models, which require a long duration or a re-activation of the central engine activity, are:

- Post energy injection into the blast wave: the faster shell is decelerated and the slower part of the outflow can catch up with it at later time injecting energy in the blast wave. This collision can generate a bump in the light curve;[10,14]
- Internal Shocks: late internal shocks produce a long duration prompt emission;[5]
- External Shock by thick shell fireballs: X-ray flares are produced in external shocks by a thick shell fireball and they mark the beginning of the afterglow emission.[1,2]

The spectral similarities observed in several bursts, between X-ray flares and afterglow emission, can be straightforwardly accounted for in the framework of External shock by thick shell fireballs and this motivated us to focus on this model.

3. X-ray flares in the framework of External Shock by thick shell fireballs

The onset of the external shock depends on the dynamical regime of the fireball, i.e. on its Lorentz factor Γ_0 and its thickness Δ, with $\Delta = ct_{eng}$ and t_{eng} the duration of the engine activity. The shell is defined to be thick if $\Delta > (E/nm_pc^2)^{1/3}\Gamma_0^{-8/3}$.[15] Writing this condition in terms of the deceleration time t_{dec} we find $\Delta > ct_{dec}$. In this case the Reverse Shock ends crossing the shell after the fireball starts to decelerate, and this implies $t_{eng} > t_{dec}$. The crossing of the RS increases the emitting volume of the shell, thus the most of the energy is transferred to the surrounding material at late time, around the end of the engine activity (Piran, private communication). If the engine releases most of the energy during the final phases of its activity, then the afterglow decay is described by a power law only if the time is measured from the instant of the central engine turns off.[16] The flare would thus be produced by an external shock caused by an energy injection lasting until the time of the flare occurrence.

3.1. *Application to several bursts*

XRR 011121 X-ray light curve shows a flare $\sim$ 240 s after the burst. The flare spectrum is softer than that of the main pulse and is consistent with the afterglow one at about 1 day.[2] The comparison of X-ray and optical data indicates that the burts occured in a wind like medium density profile. Under this assumption the light curve, from the decay part of the flare to late times, is nicely described by a power law if the origin of the time t_0 is shifted to the time of the flare, suggesting that the flare is the beginning of the afterglow emission (see Fig. 1).

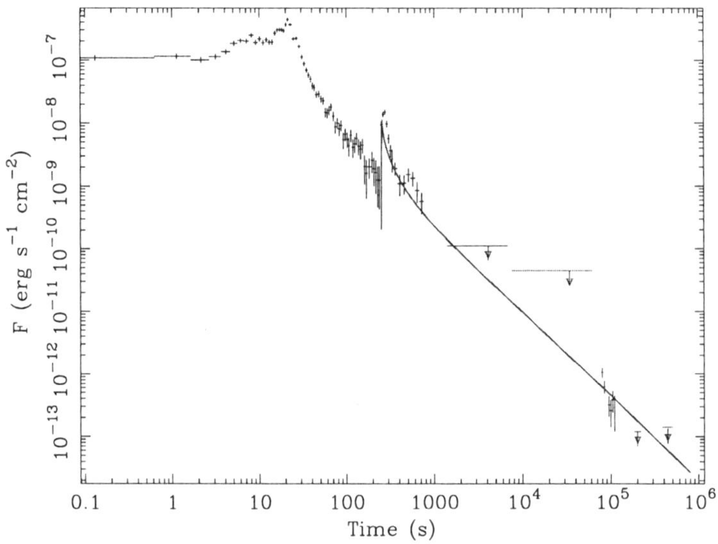

Fig. 1. X-ray light curve of XRR 011121 for a fireball expanding in a wind with the origin of the time shifted to the instant of the flare, $t_0 = 250$ s. The model parameters are $E_{53} = 0.28$, $\Gamma_0 = 130$, $A_* = 0.003$, $\epsilon_e = 0.01$, $\epsilon_B = 0.5$, and $p = 2.5$.

XRF 011030 X-ray light curve presents a flare, $\sim$ 200 s long, about 1300 s after the burst.[1] The flare spectrum is marginally softer than the main pulse and consistent with the afterglow. When t_0 is shifted to the onset of the flare the calculated light curves can describe the flare and the late data, from radio to X-ray, both in a uniform interstellar medium and in a wind profile environment (see Fig.s 12, 13, 14, 15, 16 and 17 in[1]).

GRB 050712 observed by Swift has three flares in its X-ray light curve. The third flare occurs $\sim$ 400 s after the burst and has a spectrum consistent with the following afterglow emission. The spectral ($\beta = -1.16$) and the temporal ($\alpha = -0.86$) indexes are not consistent with high latitude emission from internal shock (closure relationship $\alpha = \beta - 2$),[7] thus a thick shell fireball is required to explain the flare. Also in this case, if t_0 is shifted to the time of the flare, it connects to the afterglow with a power law (see Fig. 5 in[7]).

4. GeV flares in the contest of Late Internal Shock and External Shock model

X-ray flares overlap with the afterglow emission, thus X-ray flares photons can be Inverse Compton (IC) scattered by afterglow electrons producing flares in the GeV-TeV band. X-rayy flares appear up to thousand of senconds after the burst, thus GeV flares may be can be the cause of the delayed GeV emission detected by in EGRET in GRB 940217.

In late Internal Shock model the mechanism for X-ray flares is not clear:[17] if X-ray flares are produced by synchrotron then GeV flares are produced by self IC emission on the same electrons producing the X-ray flare, while if X-ray flares are produced by IC then GeV flares are produced by second order IC emission on the afterglow electrons.

In External Shock model by thick shell fireballs X-ray flares are produced by synchrotron thus GeV flares are produced by self-IC emission of flare photons on afterglow electrons.[18]

4.1. *Late Internal Shock*

One way to produce X-ray flares in late internal shocks is by synchrotron emission. Thus high energy flares are produced by self IC emission and X-ray and GeV flares come from the same electrons population . In the framework of late internal shock the explanation of X-ray flares requires low Lorentz factor for the shell producing the flare and a low contrast between shells Lorentz factors.[4] Low Lorentz factor contrasts imply low IC peak energies.[17] Thus, one expects a strong temporal correlation between the two flare, but not very energetic GeV flares.

Alternatively, the X-ray flare can be produced by IC emission and the high energy flare by second order IC emission. In this case X-ray flare photons need of several time to reach and scatter with the forward shock electrons, and during this time the beam spreads out. In the forward shock electrons rest frame the flare photons are anisotropic, thus the Thompson cross section decreases respect to the isotropic case, and the high energy photon emission is suppressed by a factor of several.[19] Because the angular dispersion of the up-scattered photons, the high energy flare could last much longer than the X-ray flare[20] thus one also does not expect the same temporal profile for the two flares.

4.2. *External shock*

In the framework of External Shock stronger GeV emission is expected respect that predicted in late Internal Shocks models due to higher Lorentz factors. X-ray and high energy flares are produced by the same region and electrons population, thus similar duration and temporal profiles are expected. These are strong predictions that potentially will be tested in the GLAST/Swift era.

For example, in Fig. 4.2 we show the predictions for X-ray and GeV flares for a thick shell fireball expanding in an uniform interstellar medium. The time of flare occurence is set to be t_0=500 s, and the fireball is in fast cooling regime at this time. We notice that, with the model parameters of Fig. 4.2 the GeV flare will be potentially detectable by the Large Area Telescope (LAT) on board on GLAST, which threshold is $\sim 10^{-9} erg\ cm^{-2} s^{-1} Hz^{-1}$ at 1 GeV). We present the temporal evolution of synchrotron (dashed line) and IC (solid lines) spectra in Fig. 4.2. We find that at the time of the flare appearance the peak of IC emission is between 1 and 100 GeV (green solid curve in Fig. 4.2). This causes the delay between the X-ray and the high energy flares at 1 GeV, while no temporal delay is observed at 100 GeV, where the observation frequency is above the peak of IC emission, in agreement with our expectations.

5. Conclusion

The present data suggest the existence of two families of X-ray flares, differentiated by their spectral behaviour. Spectral variablities are well explained in late Internal Shock models (but in some cases spectral variations can be explained also in the context of the External Shock). Flares with spectrum conisistent with that of the afterglow emission, instaed, can be accounted

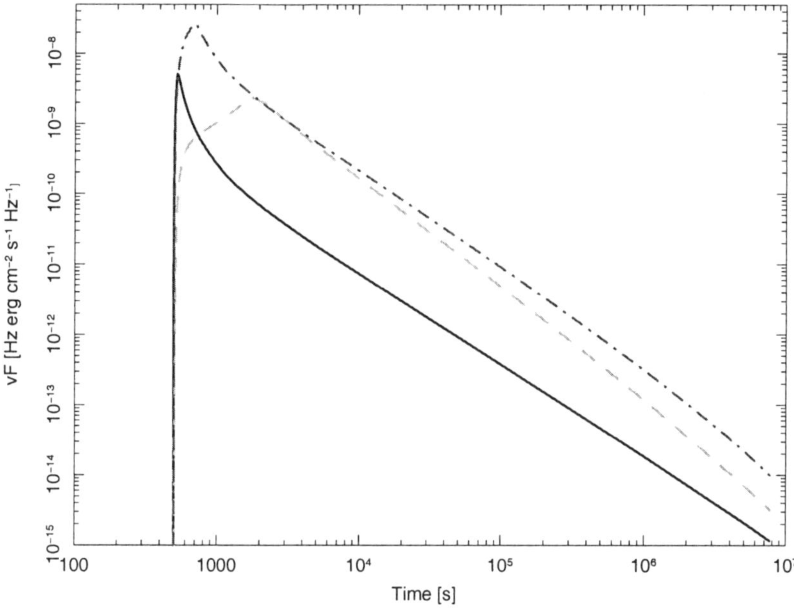

Fig. 2. High energy flares in the framework of External Shock by thick shell fireballs. The time of the X-ray flare occurance is set to be t_0 =500 s. The model parameters are Γ_0=300, E_{53}=5, n=1, ϵ_e=0.1, $\epsilon_B = 10^{-4}$, p=2.5, and z =1. The fireball is in the fast cooling regime at the time of the flare appearance. The blue dashed line is the synchrotron emission at 2 KeV. The light blue dashed line is the IC emission at 1 GeV, and the purple dashed line is the IC emission at 100 GeV.

for in the framework of External Shock by thick shell fireballs. Both in the framework of Internal Shocks and that of External Shocks X-ray flares are related to a long lasting central engine activity, and can be attended by GeV flares. In particular, in the framework of the External Shock we expect similar temporal profiles for X-ray and high energy flares. This is a strong prediction that could be tested in the GLAST-Swift era, and will permit to discriminate between the different models proposed to explain the flare phenomenology.

References

1. A. Galli & L. Piro, 2006, A&A, 455, 413G
2. L. Piro, M. De Pasquale, P. Soffitta et al., 2005, ApJ, 623, 314P
3. P.T. O'Brien, R. Willingale, J. Osborne, et al., 2006, ApJ, 647, 1213O
4. A. D. Falcone, D. N. Burrows, D. Lazzati, et al., 2006, ApJ, 611, 1005
5. D. N. Burrows, P. Romano, O. Godet, 2005, Proceedings of the conference *The X-ray Universe,astro-ph/0511039*

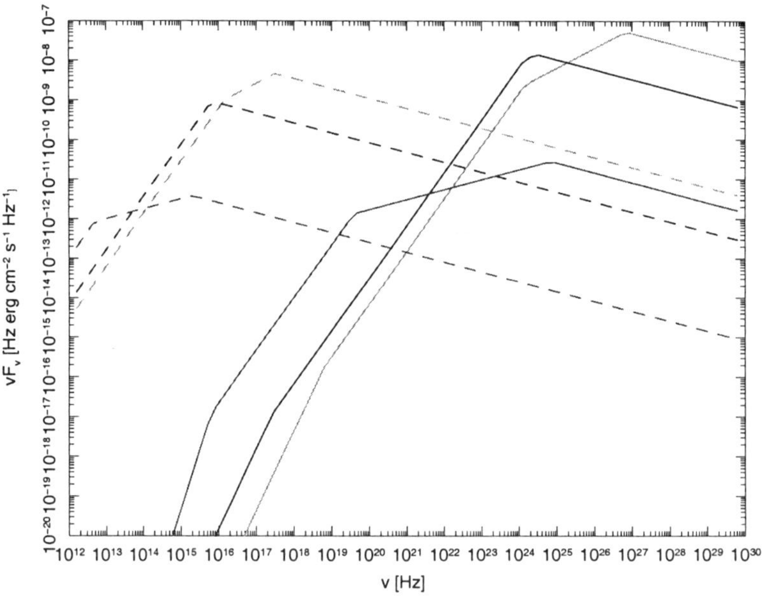

Fig. 3. Temporal evolution of synchrotron IC spectra. Solid lines represent IC emission spectra, and dashed lines synchrotron spectra. Green, blue and red lines represent the spectrum about $\sim$ 1 min, 500s and 50000 s after the flare respectively. During the first phases the fireball is in the fast cooling regime.The fast-to-slow cooling transistion takeas place between 500 s and 50000 s after the flare.

6. *G. Tagliaferri, M. Goad, G. Chincarini, et al., 2005, Nature, 436, 1132*
7. *M. De Pasquale, D. Grupe, T. S. Poole, 2006, MNRAS, 370, 1859*
8. *B. Gendre, A. Galli, A. Corsi et al., 2006, A&A submitted, astro-ph/0603*
9. *W. Zhang, S. E. Woosley, & A. Heger, 2004, ApJ, 608, 365*
10. *P. Kumar & A. Panaitescu, 2000, ApJ, 541, L9*
11. *Y. Fan & D. M. Wei, 2005, MNRAS, 364, L42*
12. *C. D. Dermer & K. E. Mittman, 1999, ApJ, 513, L5*
13. *C.D. Dermer, 2006, oral talk at the conference Swift and GRBs: Unveiling the Relativistic Universe*
14. *M. J. Rees & P. Meszaros, 1998, ApJ, 496, L1*
15. *R. Sari & T. Piran, 1999, ApJ, 520, 641*
16. *D. Lazzati & M. C. Begelman, 2006, ApJ, 641, 972*
17. *X. Wang, L. Zhuo & P. Meszaros, 2006, ApJ, 641, L89*
18. *A.Galli et al., in preparation*
19. *Y. Fan & T. Piran, 2006, MNRAS, 370, L24*
20. *A. M. Beloborodov, 2005, ApJ, 618, L13*

THE HIGHEST ENERGY EMISSION FROM GAMMA RAY BURSTS: MILAGRO'S CONSTRAINTS AND HAWC'S POTENTIAL

BRENDA DINGUS

Los Alamos National Laboratory, M.S. H803
Los Alamos, NM 87545 USA

FOR THE MILAGRO AND HAWC COLLABORATIONS

Gamma-ray bursts are amazing phenomena involving relativistic jets and the formation of black holes. Our understanding of these explosions and of the physics of their extreme environments would be greatly expanded by the observation of high energy gamma-rays. However, such observations of these unpredictable, rare, and short duration events are difficult. The Milagro detector has been used to search for gamma-ray bursts, but no significant emission has been observed. The HAWC (High Altitude Water Cherenkov) observatory is a proposed next generation detector based on the technology of Milagro that will have more than 15 times the sensitivity of Milagro to search for the highest energy emission from gamma-ray bursts.

1. Introduction

Emitting over 10^{52} ergs in gamma rays, gamma ray bursts (GRBs) are the most energetic explosions known in the universe. The emission is thought to be collimated in jets, and the bulk Lorentz factor of the particle flows may be as large as 1000. GRBs last from fractions of a second to ~1000 seconds and the duration distribution is bi-modal. The distinction between short and long bursts is roughly 2 seconds and the average duration of the short bursts is 0.2 seconds and long bursts 20 seconds. It is believed that the progenitors of short and long bursts are different. The prevailing model of short bursts is the coalescence binary neutron star systems and for long bursts the collapse of a supermassive star. In both cases the energy source is the gravitational potential energy released by the accretion of matter onto a compact object.

Theoretical considerations argue for the creation of >100 GeV gamma rays in GRBs. Shocks are created when the relativistic jet interacts with shells moving at different velocities and with the circumburst medium. Gamma rays are produced by either leptonic or hadronic processes and yield multi-wavelength spectra similar to active galactic nuclei, where the keV-MeV emission is likely due to synchrotron radiation. Measurements of the evolution of the flux at the highest energies provide strong constraints on the magnetic fields, the circumburst medium, and the bulk Lorentz factors [1].

The highest energy gamma ray conclusively detected from a GRB is an 18 GeV photon detected by EGRET roughly 90 minutes after the onset of the burst [2]. In addition, there is evidence for a high-energy component to GRBs that extends to 200 MeV with a spectral index of -1 [3]. This observation has been cited as evidence for proton acceleration in GRBs, with the implication that GRBs accelerate protons to energies above 10^{18} eV [4]. While there has been no observational evidence for a spectral break at the highest energies [5], detection of higher energy photons has proven elusive.

The rate of GRBs anywhere in the Universe is at most a few per day, and the rate at which GRBs are detected by satellites has never been greater than once per day and is currently about twice per week. Only a small fraction of these GRBs are nearby enough to observe from the ground. Gamma rays of energy E_γ emitted at a redshift z are attenuated by pair production on the extragalactic background light and the optical depth is $\tau \sim z^{4/3} (E_\gamma /90\ \mathrm{GeV})^{3/2}$ for $0.1 < z < 2$ [6]. So for a GRB at z = 0.1, 0.5 or 1, the gamma ray flux is reduced by a factor of 1/e = 0.37 at $E_\gamma \sim 700$, 170, or 90 GeV, respectively. Figure 1 shows the redshift distribution of GRBs. The short bursts are clearly closer, and ~1/3 of all bursts observed by BATSE were short bursts.

Atmospheric Cherenkov telescopes (ACTs) are able to slew to bursts within 10s of seconds, so only longest duration long bursts can be observed during the satellite detected gamma-ray emission. The early phase of long duration GRBs may have more high energy emission prior to the inevitable energy losses of the particles. Also, these first gamma rays provide the most straightforward constraints on testing for a speed of light variation with gamma-ray energy. And, the low duty cycle of ACTs of < 10 % limits the number of GRBs that can be observed.

2. Milagro's Observations of GRBs

Extensive air shower (EAS) detectors provide an alternative way to observe GRBs due to their large field of view of ~2 sr and their high duty factor of nearly 100%. The most sensitive of these detectors is Milagro—the first large, uniformly instrumented, air shower array using water Cherenkov technology. The water Cherenkov technique is an excellent method for EAS detection because of its superior detection efficiency, calorimetric capability, and low cost.

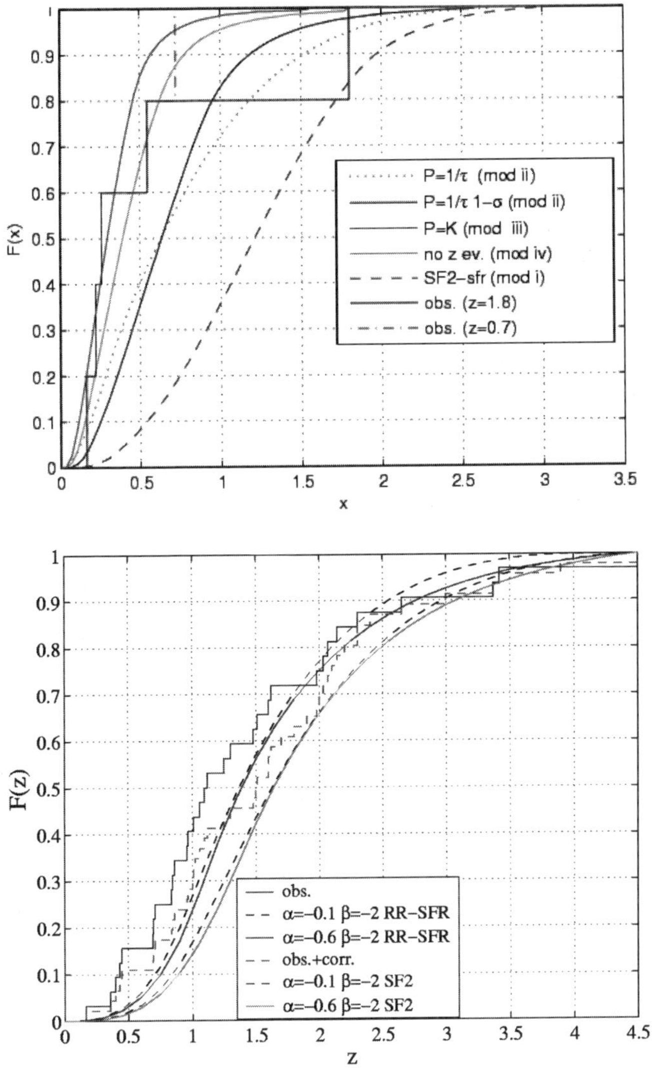

Figure 1: The redshift distribution of short, < 2 seconds, (top) and long, > 2 second duration bursts (bottom) as measured and for various models. (Top plot is from Guetta & Piran, 2006 [7] and the bottom plot is from Guetta, Piran, & Waxman, 2005 [8].) The shorter bursts are closer, but are not observable by an imaging atmospheric Cherenkov telescopes due to the slew time of the telescopes.

In contrast, gamma-ray detectors such as the Tibet ASγ [9] and the ARGO observatory [10] use a thin layer of scintillator and resistive plate chambers respectively to detect charged particles in air showers. Gamma rays, which outnumber charged particle in electromagnetic showers by 5-10 times, are only detectable with the addition of a layer of lead converter in these types of detectors. As a consequence, the ARGO detector, which is larger than Milagro and located at an elevation 1700m higher, will (upon completion) have roughly the same sensitivity as Milagro [11]. In addition, in a water detector, calorimetry of electromagnetic showers is possible, whereas thin detectors can only count particles, so Milagro can readily distinguish low energy electrons and gamma rays from muons, hadrons and high energy electromagnetic particles and use this capability to distinguish between gamma-ray (signal) and hadron (background) induced showers.

The best evidence of higher energy emission from a GRB comes from the Milagrito detector [12]. Milagrito was a smaller version of Milagro that operated for approximately one year. During Milagrito's lifetime, there were 54 bursts detected by BATSE that occurred in Milagrito's ~ 2sr field of view. One of these GRBs, 970417a, had 18 events when 3.5 were expected from background during the duration of 7.9 seconds defined by the BATSE observation. Given the trials involved in searching the large BATSE position interval and the 54 bursts searched, the chance probability of this detection is not negligible at 1.5×10^{-3}. The minimum energy of the Milagrito detected events requires this burst to have been nearer than z=0.1 in order to not be significantly attenuated by pair production with the extragalactic background.

Unfortunately, during Milagro's lifetime of ~ 6 years there has been a much lower rate satellite triggered bursts. With the launch of SWIFT, there have now been a comparable number of bursts within Milagro and Milagrito's fields of view. However, the average redshift of the SWIFT bursts is ~2.6 whereas it was ~1 for earlier detected bursts [13]. This is due to the increased sensitivity SWIFT and its smaller field of view.

During the Milagro operation no further evidence has been found of TeV emission from GRBs. However, very few bursts of low redshift have been within Milagro's field of view. Figure 2 shows the Milagro upper limit compared to the keV flux of one of the nearer and bright GRBs, 010921, that was only 10.4 degrees from Milagro's zenith [14]. The upper limits are corrected for the absorption due to pair production and two values are given for two different models of the extragalactic background light.

However, most GRBs occur without being detected by satellites. Therefore, the data of Milagro has also been searched for evidence of any astrophysical transient over time durations from a few microseconds to 2 hours. No significant excess was found. The non-observation of significant emission over any timescale can be used to set limits on TeV emission from a population of

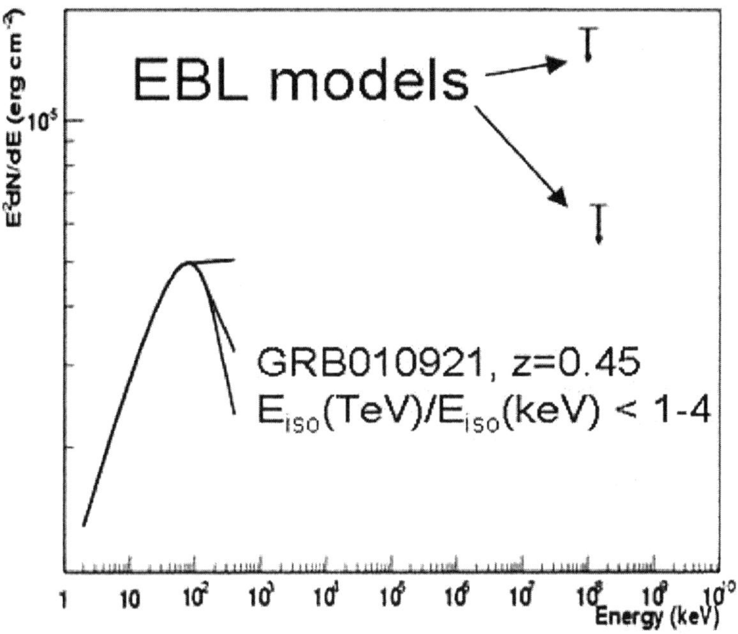

Figure 2. Milagro's upper limit on TeV emission for GRB010921 as emitted at z=0.45. Two different limits are shown for two different models of the extragalactic background light. Figure from [14].

"standard" GRBs. These limits are model dependent, depending upon the redshift distribution of the bursts, assumptions about evolutionary affects in bursts (such as correlations between intensity and or durations with redshift), and the luminosity distribution of the bursts. Under the assumptions that produce the largest TeV flux Milagro data implies that no more than 30% of all GRBs have a TeV luminosity greater than the observed keV luminosity [15].

3. HAWC's potential

The HAWC observatory combines the Milagro water Cherenkov air-shower detection technique with a very high altitude site. Re-deploying the existing Milagro photomultiplier tubes (PMTs) and electronics in a different configuration at an altitude above 4000 m will lead to a sensitivity increase of a factor of ~15 over Milagro. This dramatic improvement is due to three effects: the increased altitude, the increased physical area, and the optical isolation of the PMTs. As a result of these improvements, HAWC will detect a ~5 σ signal from the Crab Nebula in a single 4-hr transit (compared to ~5 months for Milagro). The low energy response has also been increased and is shown in Figure 3.

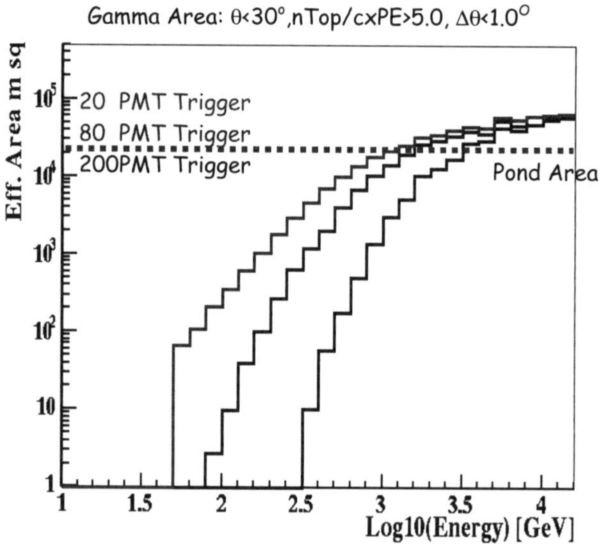

Figure 3. Effective Area of HAWC (after passing γ/hadron and angular resolution cuts for three different triggers thresholds.

The HAWC detector consists of a 150m x 150m x 5m deep reservoir lined with a polypropylene-nylon liner to contain and isolate the ~125 million liters of filtered water from the ground below. The reservoir is constructed by a combination of excavation and building up soil to form a berm around the perimeter. Milagro's 900 photomultiplier tubes will be secured on a 30x30 grid with 5m spacing. Stretching between the PMTs is an opaque curtain designed to optically isolate each sensor. With the curtains each sensor only detects light produced within its cell. This dramatically reduces the PMT noise rate, the trigger rate from single muons, and the background rejection capabilities of HAWC. The location of the HAWC detector has not yet been finalized. Suitable sites at ~ 4300m elevation have been identified at the YBJ laboratory in Tibet, China and near Sierra Negra in Mexico.

Acknowledgments

We acknowledge Scott Delay and Michael Schneider for their dedicated efforts in the construction and maintenance of the Milagro experiment. This work has been supported by the National Science Foundation (under grants PHY-0245143, -0245234, -0302000, -0400424, -0504201, ATM-0002744, the US Department of Energy (Office of High-Energy Physics and Office of Nuclear Physics), Los Alamos National Laboratory, the University of California, and the Institute of Geophysics and Planetary Physics.

References

1. Zhang, B. & Mészáros, P. 2001, ApJ, **559**,110.
2. Hurley, et al., 1994, Nature, **372**, 652.
3. Gonzalez, M. M. et al., 2003, Nature, **424**, 749.
4. Dermer, C. D. & Atoyan, A. 2004, A&A **418**, L5.
5. Dingus, B.L. 2001, High Energy Gamma-Ray Astronomy: Int. Symp. in Heidelberg, eds. F. A. Aharonian & H. J. Volk, AIP Conf. Proc, **558**, 383.
6. Kneiske, T. M., Bretz, T., Mannheim, K., Hartmann, D. H. 2004, A&A, **413**, 807.
7. Guetta & Piran, A&A, **453**, 823, 2006.
8. Guetta, Piran & Waxman, ApJ, **619**, 412, 2005
9. Amenomori, A. M. et al., 2005, ApJ **633**,1005.
10. di Sciascio, G. et al., 2005, High Energy Gamma-Ray Astronomy: 2nd Int. Symp. in Heidelberg, eds. F. A. Aharonian, H. J. Volk, & D. Horns. AIP Conf. Proc, 745. 663
11. Cao, Z. et al., 2005, 29th International Cosmic Ray Conf., Pune 5, 139.
12. Atkins, R. et al. 2000, ApJ Lett, **533**, 119.
13. Jakobsson, P. et al. 2006, astro-ph/0602071.
14. Atkins, R. et al. 2005, ApJ Lett, **630**, 996.
15. Noyes, D. 2006, Ph.D. Thesis.

Observation of GRB with MAGIC Telescope

N. Galante and A. Piccioli for the MAGIC collaboration

MAGIC is presently the imaging atmospheric Cherenkov telescope with the largest reflecting surface and the lowest energy threshold. MAGIC concluded its first year of regular observation in April 2006. During this period and the preceding commissioning phase, several Gamma-Ray Bursts were observed by the MAGIC telescope thanks to its very fast slewing capabilities. The observations were triggered by GCN alerts received from SWIFT, HETE2 and INTEGRAL satellites and were performed various minutes following the alert. The threshold energies depend on the zenith angle of the burst location and span between 80 GeV and 200 GeV . Using standard MAGIC analysis, no evidence of gamma signals was found.

Keywords: Gamma ray burst; TeV γ–ray astrophysics; Cherenkov Telescope.

1. The MAGIC Cherenkov Telescope

MAGIC (Major Atmospheric Gamma Imaging Cherenkov) telescope,[1,2] located on the Canary Island La Palma (2200 m a.s.l., 28.4′N, 17.54′W), is currently the largest imaging air Cherenkov telescope in operation. The MAGIC construction was completed in Fall 2003 and after a commissioning phase of about one year MAGIC started its first regular observation cycle in April 2005. According to detailed simulation of atmospheric showers and detector response the trigger threshold is around 60 GeV for low zenith angle observations[3] while the analysis threshold is about $E_{\mathrm{Th}} = 100$ GeV in the same conditions. MAGIC integral flux sensitivity has been calculated from Monte Carlo simulation and results in about 5% of Φ_{Crab} at $E > 100$ GeV and 2% of Φ_{Crab} at $E > 1$ TeV.[3] The angular resolution has been estimated applying the DISP method to Crab data and results to be about 0.1° for γ-ray events with $E > 200$ GeV.[4] The accuracy in point-source position determination improves as the square root of the number of collected events and is ultimately limited by tracking accuracy ($\simeq 0.02°$).[5] The energy resolution has been estimated from Monte Carlo data and results in $\Delta E/E \simeq 30\%$ at $E = 100$ GeV and $\Delta E/E \simeq 20\%$ for $E > 1$ TeV.[6] The systematic errors on the measured flux were estimated to be around

50% for the absolute flux level and 0.2 for the spectral index.

2. GRB at VHE energies

The detection of the high-energy emission from gamma-ray bursts will lead to a deeper understanding of the long standing issue of the mechanism that produce these mysterious objects. The γ-ray emission observed by the EGRET detector in some cases extends up to GeV energies.[7–9] This is in favour of an optically thin emission region and a non-thermal origin of the bursts. The observation of γ-rays at highest energies is expected to have an important impact on the modelling of the emission processes.

Very high energy (VHE) GRB observations have the potential to constrain the current GRB models on both the prompt and the extended phase of GRB emission.[10,11] Models based on either internal or external shocks predict VHE gamma-ray fluences comparable to, or in certain situations stronger than, the keV-MeV radiation, with durations ranging from shorter than the keV-MeV burst to extended TeV afterglows.[12–15]

3. MAGIC alert system

GRB are transient phenomena with duration of the order of seconds-minutes in the X-ray γ-ray energy band. Thus, being MAGIC a small field of view (FOV) instrument which can do follow-up observations and not sky-survey, a fast and efficient alert system is required. To perform a GRB follow-up observation two things are mainly required: a system which provides the GRB coordinates to follow-up instruments, i.e. which distributes the on-board calculated GRB informations by satellites to the ground based telescopes; and a system that retrieves such informations and interacts with the Telescope operators and with the Telescope subsystems too.

3.1. *The Gamma-ray burst Coordinates Network*

The *Gamma-ray burst Coordinates Network* (GCN) is a global network that provides the link between those space based instruments which perform a GRB monitoring, and the ground based instruments that require a "trigger" for the follow-up observation. It is a developement of the old *BAtse COordinates DIstribution NEtwork* (BACODINE) system, the system used by BATSE to distribute the GRB coordinates.

Figure 1 shows a schematic view of the GCN. Satellites orbiting around the Earth and monitoring GRBs must, of course, be able to acquire GRB

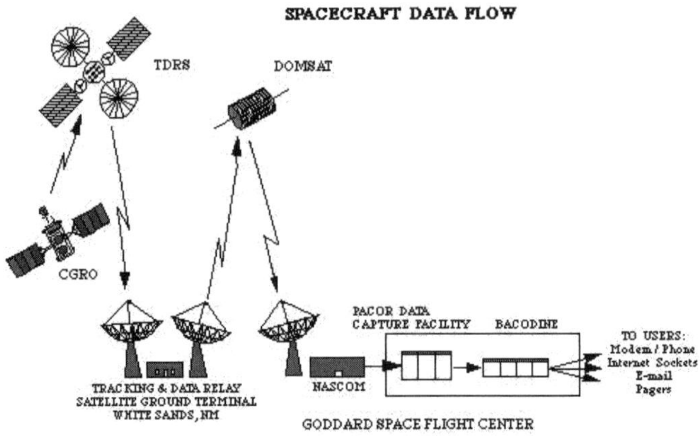

Fig. 1. Schematic view of the GCN.

data useful to calculate coordinates in short time. Then such informations are transmitted to the TDRS satellites, which can transmit all the data to the NASA White Sands Ground Station in New Mexico. From here data are retransmitted to DOMSAT and then retrasmitted again to the Goddard Space Flight Center (GSFC) Data Capture Facility. The whole data transmission procedure takes 2.048 seconds plus an additional second due to buffering between the White Sands Ground Station and the TDRS/DOMSAT satellites. Once the GRB data are at GSFC, the GRB coordinates are calculated in 0.1 seconds and then distributed, through the TCP/IP protocol, to the various ground-based instruments in 0.3 seconds.

3.2. *The MAGIC GRB Alert System*

On the other side the MAGIC Telescope must be able to catch and fastly react to the alerts distributed by the GCN.

A system called *Gamma Sources POinting Trigger* (GSPOT) has been developed: this system creates the link with GCN and manages the decoding of possible GCN alerts and the interaction with operators and with the Telescope subsystems.

Figure 2 shows a schematic view of GSPOT.

GSPOT is a daemon, a multi-thread C language program running in background and stand-alone mode at the Telescope site. It monitors 24 hours a day the GCN. In case of an alert all the informations are retrieved from the 40 long word of the GCN package thanks to the bit-masks pro-

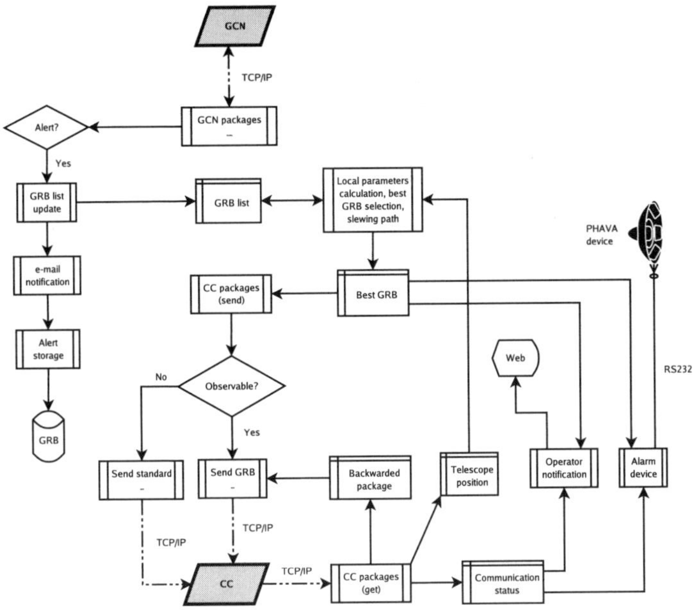

Fig. 2. Summarized flowchart of *GSPOT*.

vided by the GCN team. The alert is stored into a list, in order to keep care of all the alerts coming within a certain time, and the observability of the related GRB is computed according to some parameters defined in the GRB observation strategy. In case of observability chance, the GRB informations are re-processed and translated into human language, for notification to operators, and into MAGIC subsystems language, as long as special reports are sent to the *central control.* Thus, if the Telescope is working, an automatic repositioning procedure and GRB data aquisition can be started immediately. For safety reasons, the operator is needed to accept or not the alert.

3.3. *The GRB observation strategy*

The selection criteria are mainly affected by the Telescope possibility to observe a source at GRB local coordinates. Therefore local conditions are the main constraint to the observation by the Cherenkov technique. Several duty-cycle studies have been done,[16] providing an estimate of a duty-cycle of about 10% sr $\times$ year with the following standard duty-cycle conditions:

- **darkness**: the Sun must be below astronomical horizon;

- **wind**: wind speed must be lower than 10 m/s;
- **humidity**: relative atmosphere humidity must be less than 80%;
- **Moon**: the Moon must have at least 30° of angular distance from the GRB coordinates; in any case during the three/four days of full/almost full Moon no observation can be done.

These parameters are taken into account when selecting an alert and can be set in the *GSPOT* program. Moreover the local position of the GRB is extreemely important, being the observation by Cherenkov technique affected by the zenithal angle of the source. The maximal allowed zeith angle can be estimated requiring that the overwhelming majority of possible GRBs will have an observable spectrum. As GRBs lie at cosmological distances the gamma ray horizon affect the possible observed spectrum by applying a high energy absorption: the optical depth of the intergalactic medium is about 1 at 100 GeV for a source at a redshift $z = 1$. This means that for a source at a redshift of about $z = 1$ almost all photons above 100 GeV are absorbed by the EBL. According to this estimation of the optical depth of the extragalactic medium, the closest GRB known has a spectrum entirely absorbed above ~ 200 GeV.

The energy threshold for a IACT goes with the zenithal angle θ according to the formula

$$E_{\text{thr}}(\theta) = E_{\text{thr}}(0) \cdot \cos(\theta)^{-2.7}$$

Assuming an energy thresold of 50 GeV for the MAGIC Telescope at zenith, we estimate an energy threshold of about 11.4 TeV at 80°, 1.4 TeV at 70° and 700 GeV at 65°. Current maximal zenith values set in *GSPOT* are 55° for moonshine sky, and 60° for no moonshine sky. These are the limits considered acceptable for a GRB observation.

4. GRBs Observed by MAGIC during cycle I

Since July 15th, 2004 about 200 GRBs were detected by HETE-2, INTEGRAL and SWIFT, out of which about 100 contained GRB coordinates. Time delays to the onset of the burst were of the order of several seconds to tens of minutes. In the time period between April 2005 and March 2006 the following 11 GRBs have been observed, see table 1 by MAGIC in prompt and/or early afterglow emission phase. No signal has been found.

	GRB	Satellite	T_0 [UTC]	dT_{alert}	dT_{start}	$t_{slewing}$	zenith angle
1.	GRB050421	SWIFT	4:11:52	58 sec	83 sec	25 sec	$\sim 52°$
2.	GRB050502a	INTEGRAL	2:14:18	18 sec	990 sec	–	$\sim 30°$
3.	GRB050505	SWIFT	23:22:21	540 sec	637 sec	97 sec	$\sim 49°$
4.	GRB050509a	SWIFT	1:46:29	16 sec	35 sec	19 sec	$\sim 58°$
5.	GRB050509b	SWIFT	4:00:19	15 sec	36 sec	21 sec	$\sim 70°$
6.	GRB050528	SWIFT	4:06:45	43 sec	77 sec	34 sec	$\sim 50°$
7.	GRB050713a	SWIFT	4:29:02	13 sec	40 sec	27 sec	$\sim 49°$
8.	GRB050904	SWIFT	1:51:44	82 sec	92 sec	10 sec	$\sim 24°$
9.	GRB060121	HeteII	22:24:54	15 sec	583 sec	–	$\sim 48°$
10.	GRB060203	SWIFT	23:55:35	171 sec	185 sec	14 sec	$\sim 44°$
11.	GRB060206	SWIFT	4:46:53	16 sec	25 sec	9 sec	$\sim 13°$

5. Aknwoledgments

We would like to thank the IAC for the excellent working conditions at the ORM in La Palma. The support of the German BMBF and MPG, the Italian INFN, the Spanish CICYT, the ETH research grant TH 34/04 3, and the Polish MNiI grant 1P03D01028 is gratefully acknowledged.

References

1. Baixeras, C., et al., Nucl. Instrum. Meth. Phys. Res. A, **518**, 188 (2004)
2. Cortina, J. et al., In Proc. 29th Int. Cosmic Ray Conf. (Pune), **5**, 359 (2005)
3. P. Majumdar et al., In Proc. 29th Int. Cosmic Ray Conf. (Pune), **5**, 359 (2005)
4. E. Domingo Santamaria et al., In Proc. 29th Int. Cosmic Ray Conf. (Pune), **5**, 359 (2005)
5. B. Riegel et al., In Proc. 29th Int. Cosmic Ray Conf. (Pune), **5**, 359 (2005)
6. Wagner, R.M. et al., In Proc. 29th Int. Cosmic Ray Conf. (Pune), **4**, 163 (2005)
7. Hurley, K., et al. 1994, Nature, 372, 652
8. Dingus, B. L. 1995, Astrophysics and Space Science, 231, 187
9. Gonzalez, M. M., Dingus, B. L., Kaneko, Y., Preece, R. D., Dermer, C. D., & Briggs, M. S. 2003, Nature, 424, 749
10. Hartmann D.H., Kneiske T.M., Mannheim K.,Watanabe K., AIP Conference Proceedings, 662, 442, 2003.
11. Mannheim K., Hartmann D., Burkhardt F., ApJ, 467, 532, 1996.
12. Pilla R.P., Loeb A., ApJ, 494, L167, 1998.
13. Dermer, C. D. & Chiang J., ApJ, 537, 785, 2000.
14. Zhang B., Meszaros P., ApJ, 559, 110, 2001.
15. Razzaque S., Meszaros P., Zhang B., ApJ, 613, 1072, 2004.
16. Galante N., University of Padova, Master thesis , (available at: http://www.pd.infn.it/magic/publi.html), 2002.

GRB 060218 and the outliers with respect to the $E_p - E_{iso}$ correlation

Giancarlo Ghirlanda & Gabriele Ghisellini

Osservatorio Astronomico di Brera
Merate, I-23807, via E. Bianchi 46
E-mail: giancarlo.ghirlanda@brera.inaf.it

GRB 031203 and GRB 980425 are the two outliers with respect to the $E_p - E_{iso}$ correlation of long GRBs. These two bursts share several properties with GRB 060218, a nearby burst detected by *Swift* which instead obeys the $E_p - E_{iso}$ correlation. The SED of the prompt emission of this burst presents both thermal and non–thermal components: the former can be interpreted as due to the emission from the hot cocoon surrounding the GRB jet and the latter can be explained as synchrotron emission. During its long duration the prompt spectrum evolves from hard–to–soft and the peak energy of the time integrated spectrum occurs at ~5 keV, i.e. in the energy range of the XRT instrument. Observations at these low energies were unavailable for the two outliers, but the comparison of the available spectral informations suggests that they might be twins of GRB 060218 and, therefore, they could be only apparent outliers with respect to the $E_p - E_{iso}$ correlation. This interpretation underlines the importance of studying the broad band spectra of GRBs in order to monitor their spectral evolution throughout the entire duration of the prompt phase.

1. The peak spectral energy – isotropic energy in GRBs

Long–duration Gamma Ray Bursts (GRBs) present a correlation between the peak energy of their νF_ν spectra (E_p) and their isotropic equivalent energy (E_{iso}) emitted during the prompt phase.[1] This correlation (Fig. 1) has been updated since its discovery[1] by adding more than 38 GRBs with measured redshifts and well constrained spectral properties.[2,9,14] For a subsample of these events it was also possible to estimate their jet opening angles by measuring the jet break time of their (mainly optical) light curves. The correction of the isotropic energy for the collimation factor led to the discovery of a very tight correlation (i.e. the $E_p - E_\gamma$[9,17]) which has been used to make GRBs standard candles.[8,11] However, GRB 980425 and GRB

031203 are outliers with respect to these correlations.

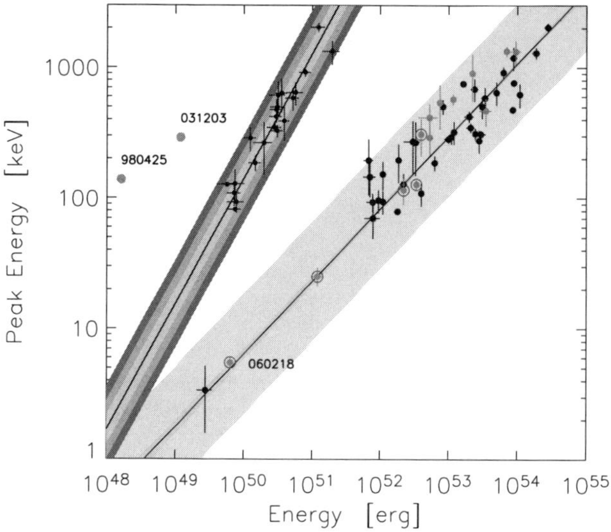

Fig. 1. Correlation between the νF_ν peak spectral energy and the isotropic energy (on the right side of the plot) defined with 49 GRBs (updated to 15 Sept. 2006). The blue points are the 15 GRBs added since 2005 (i.e. in the *Swift* "era") and the 5 events whose peak energy was measured by *Swift* are shown with red–circled blue points. The *outliers* (GRB 980425 and GRB 031203) are shown. On the left side of the plot we show the $E_{\rm p} - E_\gamma$ correlation for those GRBs with measured jet opening angle.

GRB 980425 and 031203 are associated with a nearby SN event (at $z = 0.0885$ and $z = 0.106$, respectively). However, there are at least three events which obey the $E_{\rm p} - E_{\rm iso}$ correlation and are associated with a SN event (030329, 021211 and 060218). Among these the most recently discovered (060218, Campana et al. 2006) could guide us towards the understanding of the nature of the two outliers.

It has been proposed[6,20] that the two outliers could be normal GRBs observed off axis (with a typical viewing angle twice their jet opening angle). In this scenario we can reconstruct the true energetic and peak energy of

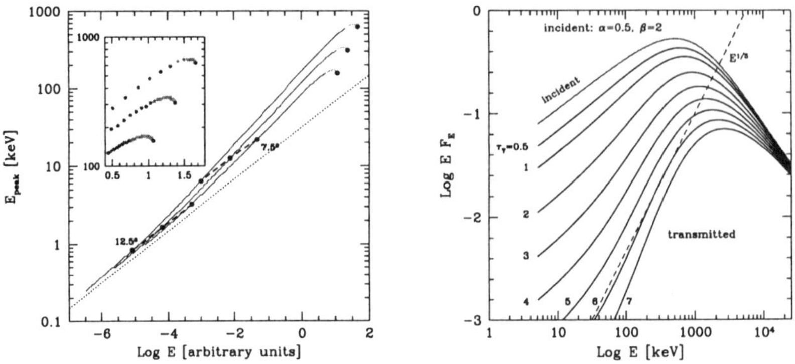

Fig. 2. **Left:** The peak energy of the observed spectrum $E_{\rm p}$ as a function of the time integrated flux E. Both depend on the viewing angle $\theta_{\rm v}$. In the insert we show a zoom for small viewing angles, within (green dots) and outside $\theta_{\rm j}$. We assumed $\theta_{\rm j} = 5^\circ$ and $0 < \theta_{\rm v} < 20^\circ$. In the comoving frame, we assume $E'_{\rm p} = 1.25$ keV. The dotted line, shown for comparison, has a 1/3 slope. The three curves are for bulk Lorentz factors $\Gamma = 50$, 100 and 200. Black points, connected with dashed lines, correspond to the same viewing angle $\theta_{\rm v} = (7.5, 12.5)$ for the three different choices of Γ. **Right:** Transmitted spectrum for different values of the Thomson optical depth $\tau_{\rm T}$, as labelled. The incident spectrum (blue solid line) peaks at $E_{\rm p} = 511$ keV, and is a smoothly broken power law, with energy spectral indices $\alpha = 0.5$ and $\beta = 2$.[12]

these two events if they were observed on axis[12] by correcting for the de–beaming effect (Fig. 2 left panel). It turns out that the two outliers should be the most luminous events in the population of bursts though being the closest (980425 is the record–holder) GRBs ever detected.

An alternative possibility is that these two bursts appear underluminous because their radiation is scattered out from the line of sight by material located in their vicinity (as proposed by[3,4,12]). The observed transmitted spectrum is different from the spectrum produced by the source (Fig. 2 right panel): for increasing optical depths the transmitted spectrum has a harder low energy component and a harder peak energy (with respect to the incident spectrum) due to the energy dependence of the Klein–Nishina scattering cross section. In this scenario the two outliers would require a screen with (Thomson) optical depths τ between 6 and 8 to become consistent with the $E_{\rm p} - E_{\rm iso}$ correlation. In this case the two outliers could have an intrinsic $E_{\rm p}$ which is consistent with the $E_{\rm p} - E_{\rm iso}$ correlation, but the presence of such a thick screen in these two bursts is problematic.

2. GRB 060218: a long burst with a peculiar SED

GRB 060218 ($z = 0.033$,[16]), associated to SN 2006aj[15] is a long duration event (>3000 s) detected by *Swift* BAT and followed with ~160 seconds delay by the XRT and UVOT telescopes onboard *Swift*.[5] Its broad band Optical to X–ray SED (Fig. 3 left panel) has some interesting features: (i) a black–body component in the X–ray with typical temperature of ~0.2 keV and a total energy of $\sim 10^{49}$ erg; (ii) a non–thermal X–ray component softening with time and (iii) an opt–UV spectrum which is well described by the Rayleigh–Jeans tail of a black body once the data are de–reddened.

The presence of a black–body component has been interpreted[5] as the SN shock breakout (SNSB) emission, which has never been observed before.

However, as shown in Fig. 2 (left panel), the opt–UV spectrum lies above the extrapolation of the X–ray black–body. Instead, a single black–body (whose Rayleigh–Jeans tail matches the opt–UV data) is inconsistent with the X–ray spectrum. Moreover, this single black–body would have a luminosity of 10^{48} erg s^{-1}. Considering the exceptional duration of this burst (i.e. $> 10^3$ s) this would imply that, if this is the energy produced by the subrelativistic SN shock breakout, it would exceed the total kinetic energy of the SN (i.e. $\sim 10^{51}$ erg) estimated from the optical spectrum.[15]

On the other hand it might still be possible that either the X–ray or the opt–UV black–body component are the SNSB. But in the first case the velocity of the emitting material $[v = L_{BB}/(4\pi t^2 \sigma T_{BB}^4)^{1/2}]$ is very small (~3000 km s^{-1}) compared to the velocity derived from the optical spectroscopy[18] (~20,000 km s^{-1}). In the second case, instead, with a peak energy of the black–body not much above the UV frequency (to limit the total energetic), the derived velocity is larger than c (if the SN exploded simultaneously to the GRB). Moreover, detailed numerical modelling of the SNSB (Li 2006) predicts lower luminosity and duration and larger temperatures for the X–ray emission of the SNSB than what observed.

We have instead proposed[13] that the opt–UV spectrum and the non–thermal simultaneous X–ray emission of GRB 060218 can be well fitted with a synchrotron model (Fig. 3 right panel) self–absorbing at a frequency just above the opt–UV band. The X–ray black–body component, instead, is the thermal emission from the hot cocoon surrounding the jet (e.g.[7,19]).

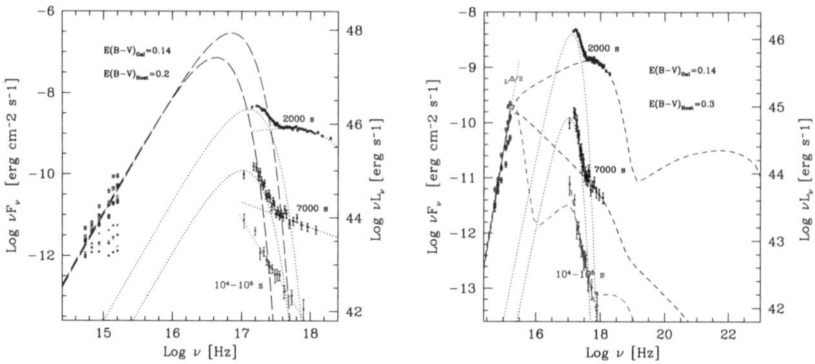

Fig. 3. **Left:** The SED of GRB 060218 at different times. Blue: 2000 s; black: 7000 s; red: 40,000 s; green: 1.2×10^5 s (only UVOT data are shown). The optical–UV data lie above the blackbody found by fitting the X–ray data (dotted lines). Instead, the opt–UV data seem to identify another black–body component (long–dashed lines) which is inconsistent with the X–ray data at the same epochs. Small crosses without error bars are UVOT data not de–absorbed. De–absorbed data [with a galactic $E(B-V) = 0.14$ plus a host $E(B-V) = 0.2$] are shown with error bars. **Right:** The SED of GRB 060218 at different times, as for the left panel, but with the optical UV points de–reddened with $E(B-V)_{\rm host} = 0.3$ instead of 0.2. This produces an opt–UV spectrum $\propto \nu^{5/2}$. Dashed lines are synchrotron self–Compton models,[13] for the 3 SEDs for which we have simultaneous UVOT, XRT data (i.e. at 2000, 7000 and $\sim 10^4$–10^5 seconds after trigger).

2.1. *The spectral evolution of GRB 060218*

One of the intriguing property of GRB 060218 is that its spectrum evolves from the hard BAT energy band to the soft XRT band. For this reason the time–integrated spectrum of this burst has a peak energy in the soft X–ray band at 5 keV. Considering its relatively low luminosity (i.e. $\sim 10^{49}$ erg – similar to that of the two outliers), its low $E_{\rm p}$ makes this burst consistent with the $E_{\rm p} - E_{\rm iso}$ correlation. In Fig. 4 we show two spectra (corresponding to the initial and the final emission of the burst) and its time integrated spectrum. The other panels show the fit with a model that reproduces the spectral evolution and the light curve in the 0.2–10 keV and 15–150 keV band and the time evolution of the $E_{\rm p}$.

The observations by the XRT instrument onboard *Swift* were crucial to measure the peak energy $E_{\rm p}$ for this burst. If we had only the BAT data, the flux would have been dominated by the bright initial phase, when the spectrum was hard. The derived $E_{\rm p}$ in this case would have been $\sim$100 keV, with a total energetics of $E_{\rm iso} \sim 7 \times 10^{48}$ erg, making this burst the third

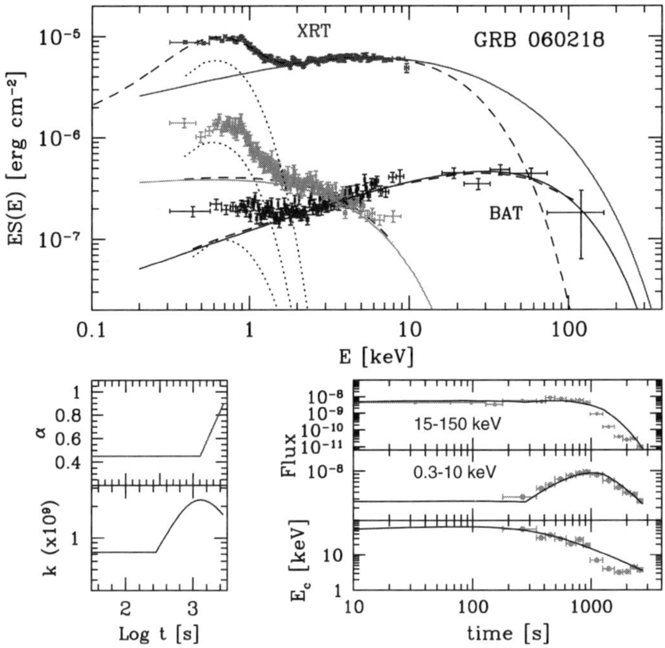

Fig. 4. Top panel: Spectra of GRB 060218 for different time–bins: i) entire duration (top); ii) [159–309 s] (rising spectrum with also BAT data) iii) [2456–2748 s] (soft spectrum). We plot $ES(E)$ vs E, $S(E)$ being the fluence. Dotted lines indicate the blackbody component, not considered for the spectral evolution, and long–dashed lines represents the best fit obtained from the analysis of the data. Continuous lines show the results of our proposed modelling. Left bottom panel: assumed behaviour of the normalisation K and energy spectral index α. Right bottom panel: light curves in the BAT (15–150 keV) and XRT (0.3–10 keV) range, and evolution of E_c. The flux in the 0.3–10 keV is the (de–absorbed) flux of the cut–off power law component only: we have subtracted the blackbody component from the total flux. Continuous lines are the results of our modelling.

outlier with respect to the $E_p - E_{iso}$ correlation.

3. Do GRB 031203 and GRB 980425 become mainstream?

We have verified if GRB 031203 and GRB 980425 can have a spectral evolution consistent with that of GRB 060218 (i.e. hard to soft). If this were

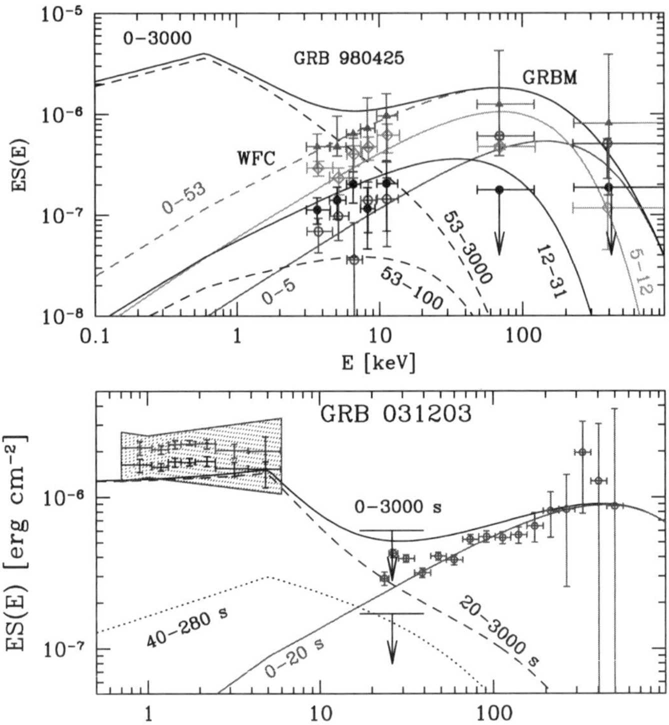

Fig. 5. *Top panel*: spectral evolution of GRB 980425. The data are from *Beppo*SAX (WFC: 2–28 keV and GRBM: 40–700 keV adapted from Frontera et al. 2000). The model fits (lines) are obtained with the same model used for GRB 060218 by simultaneously fitting the light curves and the available spectra of GRB 980425. *Bottom panel*: spectral evolution of GRB 031203. In this case the late time prompt emission should produce a considerable flux in the X–ray band below 6 keV to be consistent with the observed evolution of its dust scattering halo.[21]

the case, then it is possible that their soft late–time emission went undetected in the soft X–ray instruments on board Integral and *Beppo*SAX, which followed these two events. Interestingly, GRB 031203 (Fig. 5) produced a spectacular dust scattering halo (observed with XMM–Newton) which evolved in time. The spectral flux responsible of the halo must be at relatively soft X–ray energies (i.e. < 6 keV] and should have had a fluence similar to that of the prompt emission detected by Integral. This has two effects: first, the total energy is greater than that measured from the Integral spectrum alone while, and, second, the peak energy of the time integrated spectrum could be in the X–ray band. The combination of these two effects can make GRB 031203 consistent with the $E_{\rm p} - E_{\rm iso}$ correlation.

In the case of GRB 980425 our model requires a considerable long duration of the burst with a spectrum peaking in the soft X–ray band at late times. Unfortunately there are no data confirming this possibility (unlike the case of GRB 031203), which is however consistent with all the data we have.

Acknowledgments

We are grateful to F. Tavecchio, C. Firmani, A. Celotti and Z. Bosnjak for fruitful collaborations.

References

1. Amati, L., Frontera, F., Tavani, M., et al. 2002, A&A, 390, 81
2. Amati L. 2006, *A&A*, **372**, 233
3. Barbiellini, G. et al., *MNRAS*, **350**, L5
4. Brainerd, J.J., 1994, ApJ, 428, 21
5. Campana, S., Mangano, V., Blustin, A.J., et al., 2006, *Nature*, **442**, 1008
6. Eichler D. & Lenvinson A. 2004, *ApJ*, **614**, L13
7. Fan, Y-Z., Piran, T. & Xu, D., 2006, submitted to JCAP astro–ph/0604016
8. Firmani C. et al. 2005, *MNRAS*, **360**, L1
9. Ghirlanda G., Ghisellini G. & Lazzati D. 2004, *ApJ*, **616**, 331
10. Frontera F., et al. 2000., *ApJS*, **127**, 59
11. Ghirlanda G. et al. 2004a, *ApJ*, **613**,L13
12. Ghisellini G., et al., 2006, *MNRAS*, **372**, 1699
13. Ghisellini G, Ghirlanda G. & Tavecchio F. 2006a, *MNRAS submitted*, astro-ph/0608555
14. Lamb D. Q., Donaghy T. Q. & Graziani C. 2004 *NewAR*, **48**, 459
15. Mazzali, P. et al., 2006, *Nature*, **442**, 1018
16. Mirabal, N., et al., 2006, *ApJ*, **643**, L99
17. Nava L. et al., 2004, *A&A*, **450**, 471
18. Pian, E. et al., 2006, *Nature*, **442**, 1011
19. Ramirez–Ruiz, E., Celotti, A. & Rees, M.J., 2002. *MNRAS*, **337**, 1349
20. Ramirez–Ruiz E., 2005, *ApJ*, **625**, L91
21. Tiengo, A. & Mereghetti, S., *A&A*, **449**, 203

V Poster Session

STUDY OF THE PERFORMANCE AND CALIBRATION OF THE GLAST-LAT SILICON TRACKER

M.BRIGIDA, N.GIGLIETTO, M.N.MAZZIOTTA, P.SPINELLI
for the GLAST-LAT Collaboration

Dipartimento Interateneo di Fisica "M.Merlin" and INFN-Bari
Via Orabona 4, 70126 Bari – Italy
e-mail: monica.brigida@ba.infn.it

The GLAST Large Area Telescope (LAT) is a high-energy gamma-ray telescope that will be flown in late 2007. The LAT tracker (TKR) consists of an array of tower modules, composed of planes of silicon-strip detectors (SSDs) interleaved with converter tungsten layers. The photon detection is based on the pair conversion process and silicon-strip detectors will reconstruct tracks of electrons and positrons.
The instrument is actually assembled and the full LAT has been tested and integrated at Stanford Linear Accelerator Center (SLAC). Experimental results from the pre-launch cosmic ray data taking will be discussed.

1. The GLAST-LAT

The Large Area Telescope (LAT), the main detector onboard the GLAST satellite, will observe the gamma-ray sky in the energy range between 20 MeV and 300 GeV. It consists of a modular structure made by 16 identical towers. Each tower is composed of a hodoscopic CsI calorimeter (CAL) 8.5 radiation length thick and of a silicon tracker module made of 19 stacked trays, which hosts 18 X-Y planes equipped with SSDs for the tracking of the electron-positron pair. The array of four by four towers is shielded by segmented scintillator tiles (Anti Coincidence Detector, ACD), which provide the anticoincidence for rejection of charged particles [1,2].
The GLAST LAT simulation software implemented by the Collaboration is an Object-Oriented C++ framework called Gleam (GLAST LAT Event Analysis Machine) which allows the detailed study of the instrument performances and the development of scientific analysis software. The generation of the particles and their interaction with the detector is based on the Geant4 MonteCarlo

toolkit. The output from the simulation and the real data have the same format: the hits generated by the MC simulation are converted in digital output signal, as read by the electronics [4]. Finally, the digitized events can be processed by the reconstruction package.

2. The GLAST-LAT integrated towers

After environmental tests, TKR towers were shipped to Stanford Linear Accelerator Center (SLAC), where they have been integrated on a *grid* structure together with the CAL modules. Actually, all the sixteen towers have been assembled and tested at SLAC. Data taking with cosmic rays and Van de Graaff photons has been performed to calibrate the LAT and for a preliminary validation of the MonteCarlo (MC) simulations.[3].
Fig.1 shows the event display for a muon event from cosmic ray data taking, in the sixteen towers hardware configuration.

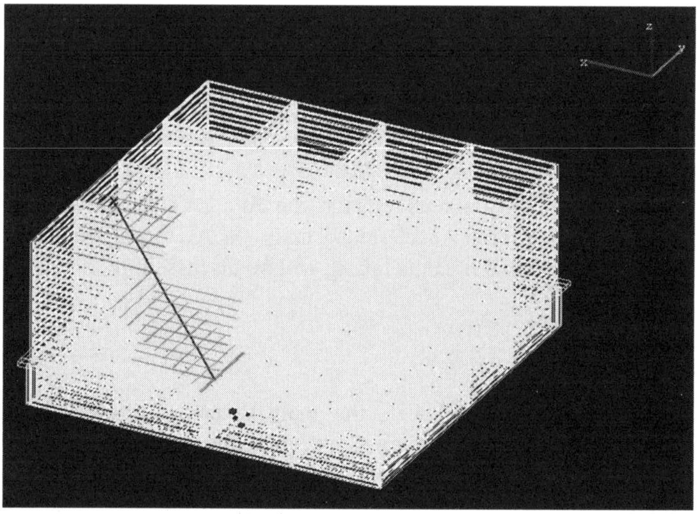

Figure 1. Event display for a muon event from cosic ray data taking

3. The cosmic ray data analysis: study of the performance of the LAT TKR

To study the performance of the LAT TKR, we have analyzed runs from cosmic ray data taking in the sixteen towers configuration, for a total of about 5×10^6 events. We selected single muons tracks fully contained in a single TKR tower.

The TKR trigger request consists of three consecutive SSD planes (both x and y views). The event selection has been made by appling the following cuts:

- events triggered by the TKR;
- only a single track;
- events contained in a tower;
- minimum ionizing particle events, selected by the CAL;

A sample of about 2×10^6 events survived to these cuts.
In order to better understand the TKR performance and to validate the MC digital output simulation we studied the hit strip multiplicity layer by layer. Fig.2 shows the average hit strip multiplicity per SSD planes for a sample of muon tracks (muon are entering from the 35th layer). The strip multiplicity is roughly constant, as expected for muon tracks, and the experimental results are well reproduced by the simulation.

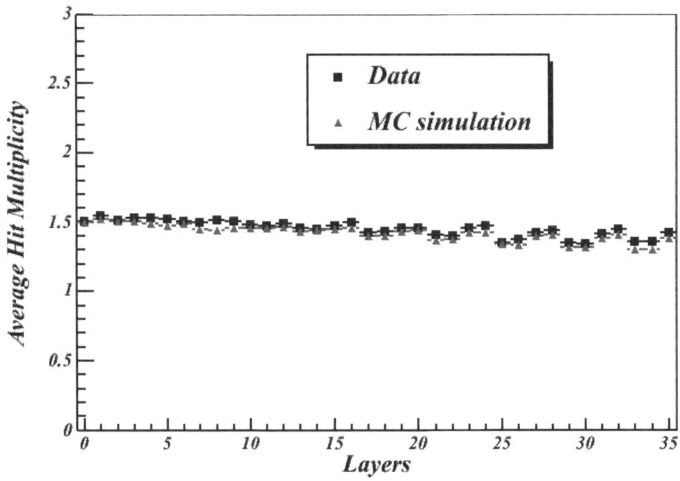

Figure 2. Hit multiplicity vs layer for cosmic ray muon data. A comparison with MC simulation is also shown.

We also have examined the dependence of the hit multiplicity on some geometrical parameters (i.e. the zenith angle θ and the azimuth angle φ).
Fig.3 shows the hit multiplicity layer by layer as a function of $1/\cos\theta$. As expected, the average number of hit strips depends linearly on the track length

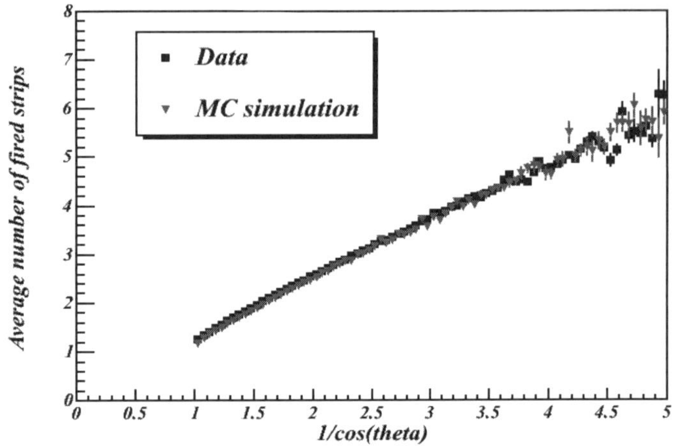

Figure 3. Hit strip multiplicity vs 1/cosθ. MC prediction are also shown.

(proportional to 1/cosθ). The simulation also reproduces the dependence of hit strip multiplicity on zenith angle.

4. Conclusions

The final results on analysis of analisys of the TRK performance in cosmic rays test in the full LAT configuration have been discussed in this paper. The study of tracker behavior shows a good agreement between data and MC simulation predictions. Results from the beam test performed during 2006 will be useful to further tune the Monte Carlo.

References

1. http://glast.gsfc.nasa.gov/
2. http://www-glast.stanford.edu/pubfiles/proposals/bigprop/
3. Eduardo do Couto e Silva and Lee Steele (2005), LAT-TD-04631-02
4. M. Brigida et al. (2002), LAT internal note, LAT-TD-0105

The Online Monitor for the GLAST Calibration Unit Beam Test

L. BALDINI, J. BREGEON* and C. SGRÒ

Istituto Nazionale di Fisica Nucleare, Sezione di Pisa,
56127 Pisa, Italy
**E-mail: johan.bregeon@pi.infn.it*

The Gamma-ray Large Area Space Telescope (**GLAST**[1]) will be launched in autumn 2007 to observe the γ-ray sky from $20MeV$ to $300GeV$. The spare flight modules, integrated in a Calibration Unit (CU), underwent a beam test campaign at CERN during summer 2006. An online monitoring software was specifically developed to check for hardware performance and beam features.

Keywords: GLAST; Beam test; Online monitoring; Calibration Unit

Introduction

GLAST is a new generation space telescope dedicated to γ-ray astronomy. The Large Area Telescope (**LAT**), the main instrument on-board the satellite, is a γ-ray pair conversion telescope built as a 4×4 grid of identical modules, each one consisting of a micro-strip silicon tracker and a CsI hodoscopic calorimeter.

During summer 2006, a beamtest campaign took place at CERN PS and SPS with two main goals: benchmark the Monte-Carlo simulation, and check the instrument performance. Here we describe the `Online Monitor` (**OM**), a software tool devoted to control the detectors response and basic beam properties in real time.

1. The Calibration Unit beam test

The **CU**, integrated for the beam test purpose, is composed of two tracker towers and three calorimeter modules enclosed in an Al grid container. Five spare tiles of scintillator from the LAT AntiCoincidence Detector (**ACD**), placed outside the container, complete the assembly.

The experiment at CERN PS was mainly dedicated to the measurement of tagged γ-rays (from 50 MeV to 1.5 GeV) and electromagnetic and hadronic cascades (from 2.5 GeV/c to 10 GeV/c). Photons were tagged using a set of Ancillary Detectors (**AD**) readout by a separate DAQ system; the synchronization with the CU DAQ has been a key point of the experiment.

At CERN SPS, the CU response was studied with very high energy electrons and hadrons (from 10 GeV/c up to 280 GeV/c).

2. The Online Monitor

As the offline data processing cannot be completed within less than a couple of hours, a tool to monitor in real time the behavior of the instrument and the beam quality was of crucial importance.

2.1. *Architecture*

The `Online Monitor` relies on a feature of the GLAST ground operation DAQ, which is able to multicast data packets over a local network, as they are acquired. Hence, the OM can be run on different computers without any impact on the DAQ.

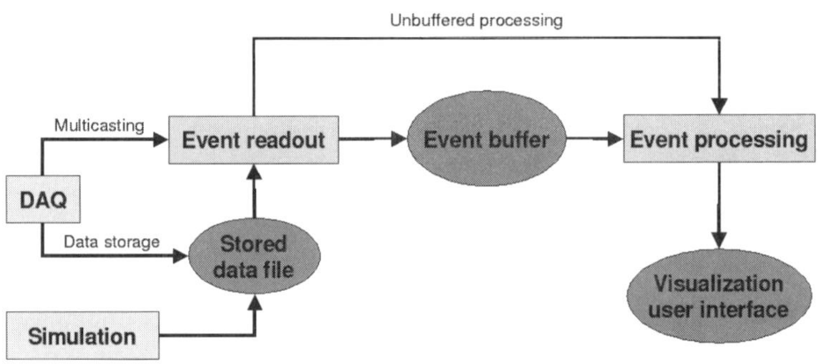

Fig. 1. Conceptual architecture of the `Online Monitor`.

The software is built around a `Python`[2] core with `Hippodraw`[3] as a data analysis and visualization module, and `Qt`[4] for the Graphical User Interface. The performance, in terms of processing efficiency, was greatly improved by implementing a multithreaded structure (one thread gets data while another processes events), coupled with a self-adapting dynamic event buffering (Fig.1). Thanks to its modular and flexible architecture, the software could constantly evolve to match many different hardware setup. The

OM also features a read back capability which has been extensively used on raw data files, during the beam test, and on simulated data during its development.

2.2. *Monitored quantities*

Fig.2 shows some of the distributions (beam position and angle in the CU reference frame, raw energy deposit in the calorimeter) as they are displayed on the OM main tab. It is worth noticing that quick online reconstruction algorithms provide an accuracy on these quantities which is typically within a factor of two worse with respect to the standard offline analysis.

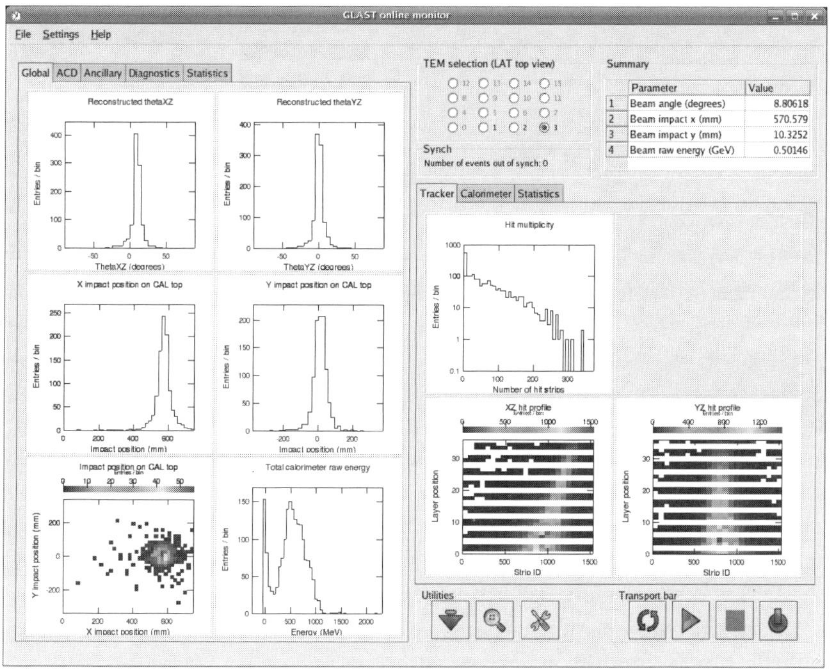

Fig. 2. The main (left) and tracker (right) panels of the OM (real data, tagged photons).

The OM also provides some basic distributions (number of hits, bi-dimensional hit maps) related to the single tracker and calorimeter modules, as well as the pulse height measured by the ACD tiles. Other features include the monitor of the AD (Fig.3) and the event by event check of the synchronization between the two DAQs.

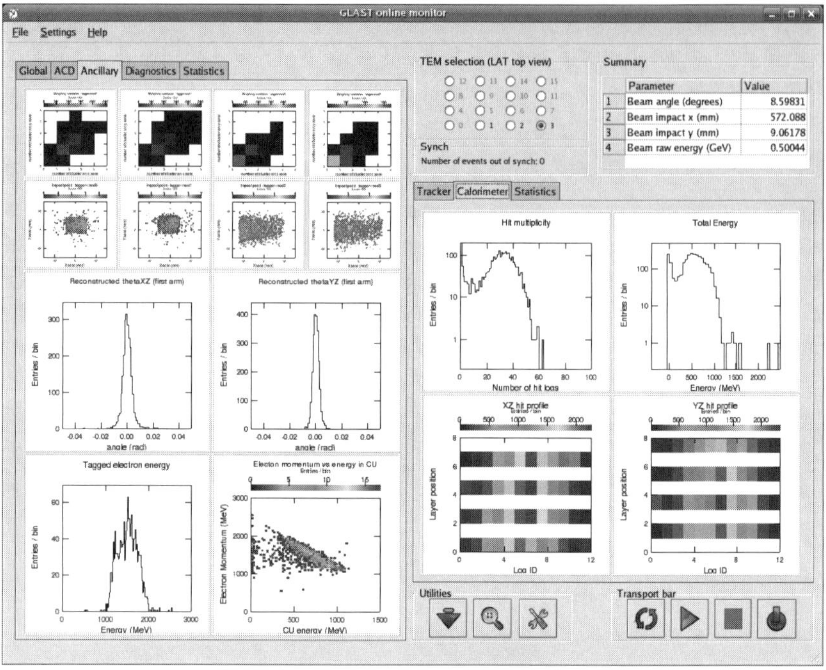

Fig. 3. The AD (left) and calorimeter (right) panels of the OM (real data, tagged photons).

Conclusion

For the purpose of the GLAST LAT Calibration Unit beam test at CERN, we developed a software to monitor online the beam properties and the detector response. Getting events multicasted on the network, the OM was run in parallel on three computers without having any impact on the DAQ. The OM was successfuly used to verify the position of the CU with respect to the beam while acquiring data, calibrate the photon tagger as well as confirm in real time the synchronization between data streams.

References

1. J. E. Grove, J. E. McEnery, GLAST LAT Collaboration, *GLAST Large Area Telescope Instrument Integration and Science Preparation*, Bulletin of the American Astronomical Society, Vol. 37, p.1198
2. http://www.python.org
3. http://www.slac.stanford.edu/grp/ek/hippodraw
4. http://www.trolltech.com/products/qt

ARGO-YBJ experiment: the Scaler Mode Technique

Irina James* on behalf of ARGO-YBJ collaboration

Department of Nuclear and Theoretical Physics, Pavia University,
27100 Pavia, Italy
**E-mail: irina.james@pv.infn.it*
http://argo.na.infn.it

The ARGO-YBJ detector is an Extensive Air Shower (EAS) array located in Tibet (P.R. China) at 4300 m a.s.l., performing a continuos sky observation by shower reconstruction with an energy threshold of a few hundreds of GeV. To lower this energy threshold to E$\sim$1 GeV the detector has been designed to work in scaler mode, i.e. recording the counting rate of each module of the detector at fixed time intervals. At these energies, signals due to local (e.g. solar events) and cosmological (e.g. Gamma Ray Bursts, GRBs) phenomena are expected as a significant enhancement of the counting rate over the background. Results on the search for GRBs in coincidence with satellite detections are presented.

Keywords: Cosmic Rays, Gamma Ray Bursts, Resistive Plate Chambers, Extensive Air Showers

1. Introduction

Emission from GRBs in the GeV-TeV energy range has been predicted theoretically and the possible detection of this high energy component plays an important role in constraning different emission models. Emission above 1 GeV has been reported for 3 GRBs, with one photon reaching 18 GeV [1], proving that the spectra of at least some GRBs extend to the GeV energy range. The ARGO-YBJ experiment, thanks to the large field of view (> 4 sr, limited only by the geometrical acceptance), the high duty cycle and the low energy threshold (especially in scaler mode), is particularly suitable to search for these transient phenomena.

2. The detector

The detector consists of a single layer of Resistive Plate Chambers (RPC) operating in streamer mode and has a modular structure, the basic module being the cluster (5.7×7.6 m^2) (see [2] and references quoted therein, for a complete apparatus description). The whole carpet will be made of 154 clusters, of which 130 are in data taking, with a detection area of ~6700 m^2 and ~93% of active area. The detector is connected to two different DAQs, corresponding to two operation modes: the shower and the scaler modes. In shower mode, for each event the location and timing of the secondary particles is recorded, allowing the reconstruction of the lateral distribution and the arrival direction with a threshold energy of a few hundreds of GeV. The lower limit of the detector (E~1 GeV) is reached by the scaler mode technique, in which the total counting rates of each cluster are recorded every 500 ms, with no information on the arrival direction and spatial distribution of the detected particles. Four low multiplicities channels in each cluster are implemented for event multiplicities from ≥ 1 to ≥ 4. The measured counting rates are respectively $\sim$ 40 KHz, $\sim$ 2 KHz, $\sim$ 300 Hz and $\sim$ 120 Hz. The counting rate for a given multiplicity is obtained using the following relation: $n_i = n_{\geq i} - n_{\geq i+1}$, where n_i is the counting rate for the i_{th} multiplicity.

Dealing with transient phenomena, it is important to understand the detector statistical behaviour. For each Cluster, we studied the counting rate distribution for each multiplicity. As shown in figure 1a, the exact multiplicity counting rate follows a Poissonian distribution, while the total counting rate added up on the 4 multiplicities channels (figure 1b) follows, due to correlation between different scalers, a distribution which is larger than a Poissonian distribution, with a σ^2 given by the following expression:

$$\sigma^2(n_{tot}) = \sigma^2(n_1) + 4\sigma^2(n_2) + 9\sigma^2(n_3) + 16\sigma^2(n_4).$$

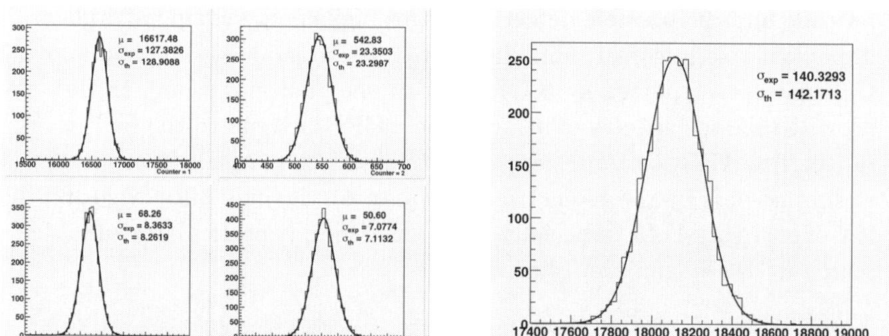

Fig. 1. a. Counting distribution for each multiplicity over a period of 30 minutes for a typical cluster; b. Total counting rate summed over the multiplicities channels for the same cluster.

In order to estimate the sensitivity of the detector, the effective area of the complete detector has been calculated for each multiplicity with a MonteCarlo simulation for photons and protons in the energy range between 1GeV and 1TeV and at zenith angle $\theta = 0°, 10°, 20°, 40°$ (see figure 2). For multiplicity n = 1 the detector sensitivity does not depend on its geometrical features and the effective areas for any carpet dimension can be scaled from the plotted values [3].

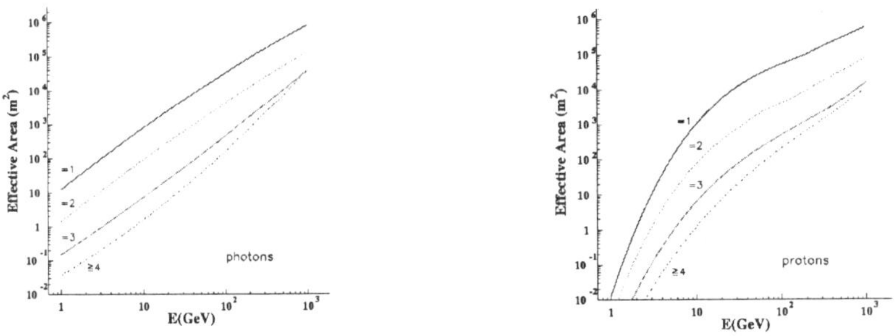

Fig. 2. Effective areas for primary photons (left plot) and protons (right plot) with zenith angle $\theta = 20°$. The curves refer to different multiplicity channels: n = 1, n = 2, n = 3 and n $\geq$ 4.

3. Search for high energy emission from GRB in coincidence with satellite

Following the first GRB detection by Swift on December 17, 2004, a search has been conducted for an excess in the counting rate of the single multiplicity channel during the time duration Δt_{90} measured by the satellites, estimating the background from the average counting rate in a time window 20 times the Δt_{90}. Up to May 2006, 28 GRBs detected by satellites were within $\theta \leq 40°$, but due to the detector installation and debugging operations, reliable data are available only for 16 of them. For each GRB the excess of the signal in units of standard deviations has been calculated and it is shown in column 8 in table 1. In particular three GRBs (GRB05114, GRB060105 and GRB060510A) show a statistical significance equal to 2.8, 3.6 and 3.7 respectively. Taking into account that we considered a sample of 16 GRBs, these values correspond to a post-trial probability of 1.7σ, 2.8σ and 2.9σ respectively. For none of these 16 GRBs a significant excess has been observed in the burst time window and 3σ fluence upper limits were set in the energy range 1 - 100 GeV using the spectral index measured by the satellites. For GRBs with known redshift, the upper limit was calculated assuming the $\gamma\gamma$ absorption model of Kneiske et al [4].

Table 1. List of GRBs searched by ARGO-YBJ with preliminary upper limits

GRB	Instrum.	T90 (s)	θ (deg)	Redshift	Spectral Index	Carpet Area (m^2)	σ	Fluence UL (erg/cm^2)
041228	Swift	62	28.1	–	1.56	693	-1.3	$3.3\ 10^{-4}$
050408	HETE	15	20.4	1.24	1.98	1820	-2.2	$9.6\ 10^{-5}$
050509A	Swift	12	34.0	–	2.1	1820	0.29	$1.6\ 10^{-4}$
050528	Swift	11	37.8	–	2.3	1820	-0.012	$6.5\ 10^{-4}$
050802	Swift	20	22.5	1.71	1.55	1820	0.74	$1.0\ 10^{-4}$
051022	HETE	200	37.9	0.8	1.22	3379	-0.68	$9.8\ 10^{-4}$
051105A	Swift	0.3	28.5	–	1.33	3379	0.90	$1.4\ 10^{-5}$
051114	Swift	2	32.8	–	1.22	3379	2.8	$1.9\ 10^{-5}$
051227	Swift	8	22.8	–	1.31	3379	0.93	$2.5\ 10^{-5}$
060105	Swift	55	16.3	–	1.11	3379	3.6	$5.9\ 10^{-5}$
060111	Swift	13	10.8	–	1.63	3379	0.82	$2.5\ 10^{-5}$
060115	Swift	142	16.6	3.53	1.76	4505	-2.2	$2.3\ 10^{-4}$
060421	Swift	11	39.3	–	1.53	4505	-0.46	$1.6\ 10^{-4}$
060424	Swift	37	6.7	–	1.72	4505	1.9	$4.1\ 10^{-5}$
060427	Swift	64	32.6	–	1.87	4505	-1.8	$1.8\ 10^{-4}$
060510A	Swift	21	37.4	–	1.55	4505	3.7	$2.3\ 10^{-4}$
060526	Swift	14	31.7	3.21	1.66	4505	0.75	$1.2\ 10^{-4}$

4. Conclusion

The ARGO-YBJ detector installation is almost completed and the detector is already in data taking, showing the expected statistical behaviour. Until now, the search for high energy emission from GRBs has given no positive result. The scaler mode has already shown a good sensitivity, with fluence upper limits $\sim 10^{-4}$ - 10^{-5} erg/cm^2 in the energy range between 1 - 100 GeV.

References

1. J.R. Catelli et al., AIP Conf. Proc. No.428, p.309 (AIP New York 1998).
2. 29th ICRC Pune, India (August 3 - August 10, 2005).
3. S. Vernetto, Astrop. Phys. 13 (2000) 75.
4. T.M. Kneiske et al. A&A 413 (2004) 807.

Analysis of Pulsars in LAT Data Challenge 2: a population point of view

M. RAZZANO

Universitá di Pisa and Istituto Nazionale di Fisica Nucleare Sezione di Pisa, Largo B. Pontecorvo 3, 56127 Pisa, Italy
**E-mail: massimiliano.razzano@pi.infn.it*

One of the milestones in the preparation of the software of the Gamma-ray Large Area Space Telescope (GLAST) was the second Data Challenge (DC2). The results of the analysis of simulated pulsars in DC2 is presented together with the description of the automated procedure used in the analysis.

Keywords: Pulsars, Gamma-rays, GLAST

1. The GLAST LAT Data Challenge 2

Pulsars are among the best target of the Gamma-ray Large Area Telescope (GLAST), an international space mission entirely devoted to the study of the gamma-ray Universe. GLAST is scheduled for launch in 2007 and carries two main instruments, the Large Area Telescope (LAT) and the GLAST Burst Monitor (GBM). The LAT is the main GLAST instrument and consists of a pair conversion telescope that can detect gamma-rays from about 30 MeV up to 300 GeV. The LAT will observe the gamma-ray sky with unprecedented detail thanks to its superior resolution and sensitivity and will cover the spectral gap existing between the space-based missions and the Air Cherenkov Telescopes on the ground, unveiling the cosmos in the energy band from about 30 GeV up to 100 GeV, today unknown.In order to validate LAT MonteCarlo, to test the LAT analysis tools and to study the LAT Instrument Response Functions (IRFs) and science capabilities, the LAT collaboration has planned a series of *Data Challenges*. The second Data Challenge (DC2) has been conducted during the period from March to the end of May 2006. During DC2 LAT scientists analyzed a set of simulated LAT data covering an observation of 55 days in scanning mode. In order to gain more confidence with the data and to better exercise the tools, the collaboration developed a quite realistic model of the sky containing the known gamma-ray sources and adding new fake sources and possible candidate sources of gamma-rays. A picture of such simulated sky is in Fig.1. Each Science Working Group in the LAT Collaboration worked on a specific class of sources, e.g. blazars, GRBs or pulsars. The pulsars simulated in DC2 were obtained using the *PulsarSpectrum*[1] simulator, a program that can simulate with high detail the gamma-ray emission from pulsars. The population of pulsars in DC2 included the six known EGRET pulsars, i.e. Vela, Crab, Geminga, PSR B1706-44, PSR B1055-52 and PSR B1951+32.[2] In addition a set of 39 fake gamma-ray pulsars coincident with radio pulsars within 3EG sources were added. An additional population of 140 pulsars with a spectrum according to the Slot Gap scenario were included and a component of 229 millisecond pulsars was also added. For some of these

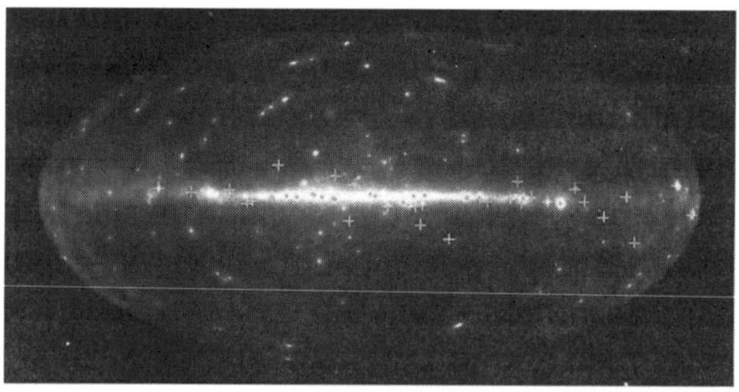

Fig. 1. A count map showing the DC2 simulated sky. The circles represent the LAT sources recognized using the maximum of likelihood spatial analysis and the crosses indicate the simulated pulsars within the LAT sources error bar.

simulated pulsars a correspondent set of radio ephemerides were provided, while for the others the ephemerides were keep hidden, simulating the situation of Radio-Quiet pulsars (RQ), useful for the LAT users that want exercise blind searches techniques. In this paper a study on the population of Radio-Loud (RL) pulsars, i.e. with a radio counterpart is presented.

2. Analysis Tools and procedures

In order to analyze pulsars in DC2 a suite of Python scripts called *pyPulsar* was designed. These scripts uses the LAT Science Analysis Environment(SAE) to analyze the LAT DC2 data, performing some analysis steps. First of all a selection of the photons was performed. Since the LAT Point Spread Function (PSF) is energy-dependent, an optimal cut would be to select photons coming from an energy-dependent radius around the source. In this first analysis we used instead a fixed radius, analyzing the pulsars with two energy cuts. First we selected photons of energies above 100 MeV within a radius of 3° and then for energies above 1 GeV within a radius of 0.5°. The values of the radii were chosen in order to have a radius comparable with the 68% of cointainment. After barycentering the periodicity was tested using the χ^2-Test,[3] implemented in the SAE. We then set some chance probability P threshold in order to define a low-confindence detection ($10^{-9} < P < 2 \times 10^{-3}$) and high-confidence detection ($P < 10^{-9}$). These values were taken in agreement with the value used for EGRET pulsars.[2]

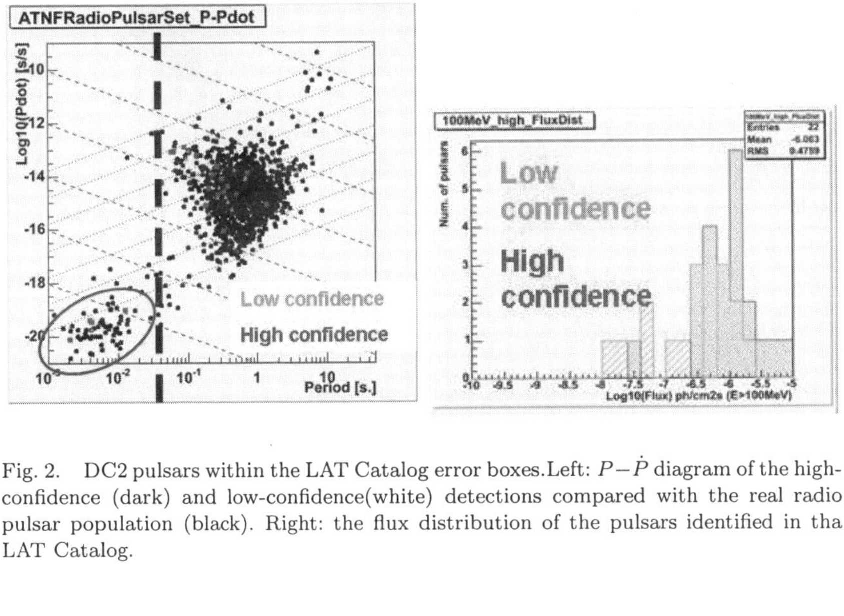

Fig. 2. DC2 pulsars within the LAT Catalog error boxes.Left: $P-\dot{P}$ diagram of the high-confidence (dark) and low-confidence(white) detections compared with the real radio pulsar population (black). Right: the flux distribution of the pulsars identified in tha LAT Catalog.

3. Results from the DC2 pulsar population

The first step of this analysis was the search for gamma-ray pulsars within the LAT DC2 catalog. In fact part of the DC2 analysis consisted in the creation of a catalog of gamma-ray sources using some techniques, among others the maximum of likelihood technique as for the EGRET data. In these data 22 high-confidence detections and 13 low-confidence detections were found, among others a millisecond pulsar (Fig. 2). For low-confidence detections a lower flux of about $10^{-8} phcm^{-2}s^{-1}$ is shown, about 1 order of magnitude lower than the faintest of EGRET pulsars. The second step in the analysis was devoted to the search of pulsars that were not in the LAT Catalog but that could be detected by using the time signature using the timing solutions provided in the DC2 pulsar database. A separate search for energies above 100 MeV and 1 GeV was performed, leading to 32 high-confidence and 51 low-confidence detections, as shown in Fig3.

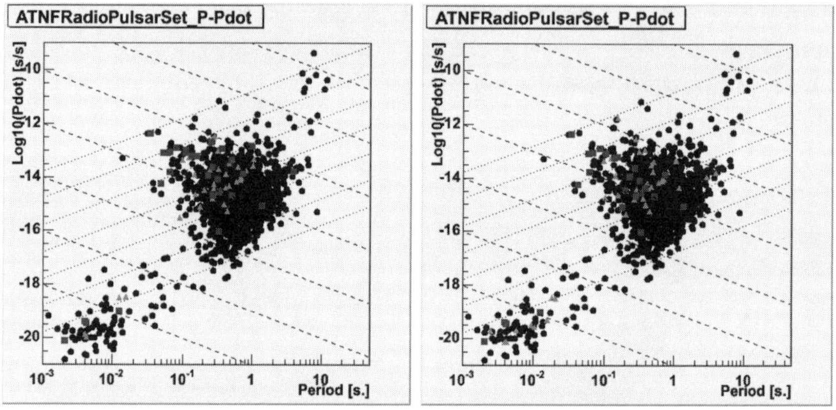

Fig. 3. Pulsars not in the LAT Catalog and identified using the timing solutions. Left: using gamma-rays above 100 MeV. Right: Using gamma-rays above 1 GeV. The color scheme is the same of Fig.2.

4. Conclusions

The analysis presented here is a basic one that used the simulated data of the LAT Data Challenge 2. From this analysis the feasibility of an automated procedure is presented and some first results on the LAT capabilities for pulsar are shown, confirming the great discovery potential for pulsars of the future GLAST mission.

References

1. M. Razzano, L. Latronico, N. Omodei and G. Spandre, *ArXiv Astrophysics e-prints* (Octtober 2005).
2. D. J. Thompson, Gamma-ray Pulsars: Observations, in *American Institute of Physics Conference Series*, 2001.
3. D. A. Leahy, W. Darbro, R. F. Elsner, M. C. Weisskopf, S. Kahn, P. G. Sutherland and J. E. Grindlay, *The Astrophysical Journal* **266**, 160(March 1983).

SEARCH OF OPTIMIZED CUTS FOR GAMMA-RAY PULSAR DETECTION WITH GLAST-LAT INSTRUMENT

G. A. CALIANDRO[†], N. GIGLIETTO, P. SPINELLI

Dipartimento interateneo di Fisica"M.Merlin" and INFN–Bari, Via Orabona 4, Bari-Italy

In view of the upcoming launch of new generation gamma-ray satellites, new analysis methods for pulsar detection are in development. We have formulated a new kind of period test that self contain the photon selection cut optimization for each single pulsar. Comparing these tests with the already known period tests, it results in an overall increase in power of ~20% with respect to the *Z test* and ~50% with respect to the *H test*.

1. Introduction

In 2007 the first satellites of the new generation gamma-ray telescopes are planned to be launched. A new era of discoveries in the gamma-ray sky will be opened. One of the most important topics will be the study and the detection of new gamma-ray pulsars.

This work concerns the detection of pulsed emission from radio-loud gamma-ray pulsars with the *Large Area Telescope* (LAT) on board the *Gamma-ray Large Area Space Telescope* (GLAST).

In gamma-ray data, the very low rate of photons emitted by the sources and the orbital motions of the satellite cause the destruction of the signal coherence, especially for pulsed emission with short period like that of pulsars. Therefore the detection of pulsed emission in gamma-ray energy band is not trivial and in general the knowledge a priori of the pulsar ephemeris is very helpful. This condition is satisfied in the case of gamma-ray counterpart detection of radio-loud pulsars.

A fundamental question is the determination of the confidence level for the detection of pulsed emission. With this aim several tests for periodicity were developed. The most used are the χ^2 *test* (Leahy et al. 1983a,b), the Z_m^2 *test* (Buccheri et al. 1983) and the *H test* (De Jager et al. 1989). All these tests for periodicity are statistical tests for the null hypothesis. In the case of a pulsar, the null hypothesis means no pulsation: i.e. a uniform distribution in the phase

[†] andrea.caliandro@ba.infn.it

histogram. So a period test returns the probability that a non-uniform phase histogram is due to statistical fluctuations of data.

Of course a period test is to be applied to a sample of photons selected around the source. The photon selection criteria affect the result of the period test and it could be a discriminating factor in detecting pulsed emission, especially in the case of weak pulsars.

We have investigated the photon selection criteria to apply to gamma-ray data in order to obtain the best confidence level for the pulsed emission.

This investigation has led to the formulation of a new kind of period test that self contains the optimization of the photon selection criteria, taking implicitly into account the characteristics of the instrument and the features of the gamma-ray source.

This new kind of period test is essentially an improvement of the old tests and it will be described in §2. In §3 are summarized the results obtained applying this improvement to some simulations and the comparison with the old tests. Finally a brief discussion will follow in §4.

2. Improved test

In analysing a point-like source, the primary photon selection criteria to apply regard the choice of the Region Of Interest (ROI), the energy band selection and the temporal selection. The ROI is usually defined as a disk centred on the source. Concerning the temporal selection for periodic sources, like pulsars, it is always useful to analyse as long a period of time as possible; it will be the entire time of observation unless a glitch happens.

The choice of the best photon selection criteria to detect pulsed emission is related to many factors. It depends on the source features, diffuse emission local properties and on the instrument performance. Therefore it is not possible to define omnibus criteria to apply to all the pulsars for which a period test give the maximum confidence level for the pulsed emission.

To overcome this problem we have formulated an improvement for the period tests. The improvement consists in performing a period test for several different choices of photon selections and taking the maximum confidence level among the values obtained.

We have applied this improvement with the *H test* and the Z_m^2 *test* with m=2 and have called it respectively *H improved test* and *Z improved test*.

The set of photon selections to take into account is to be chosen a priori. We have considered all the pairs *[ROI radius, Energy band]* given by the combination of the following cuts:

– *ROI radius: 0.25°, 0.50°, 0.75°, 1°, 1.5°, 2°, 3°, 4°, 6°, 8°*

– *Energy band: E>20MeV, E>50MeV, E>100MeV, E>200MeV, E>400MeV, E>600MeV, E>1GeV, E>10GeV*

for a total of 80 photon selection criteria.

Concerning the statistic point of view, the introduction of these 80 trials of photon selection criteria is expected to produce a distortion in the distribution under the null hypothesis of the original period tests.

In order to obtain the new distribution, we have applied the improved tests on a sample of 4500 spurious sources. For this analysis a spurious source has been defined as a random point in the DC2 sky, excluding all the pulsars.

The results are shown in figures 1 and 2 for the *H improved test* and *Z improved test* respectively. As expected the two distributions strongly differ from the exponential distribution of the original *H test* and from the χ^2 distribution of the *Z test*.

These results can be explained taking into account that the distribution under null hypothesis of the *improved tests* is strongly affected by the definition and the number of the photon selection criteria as well as by the instrument characteristics.

3. Application

In 2006 the GLAST team collaboration organized a detailed simulation called *Data Challenge 2* (DC2). The *DC2* consists in a simulation of 55 days of the high-energy gamma-ray sky observed with *GLAST-LAT* instrument.

We analysed all the pulsars in the *DC2* sky with the *H* and *Z* period test and with their relative *improved tests*. The *H* and *Z tests* were applied with an *ROI* of radius 3° and energy *E*>100*MeV*. We have considered as detected all the pulsars whose pulsed emission has a confidence level better than 99.9%.

Comparing the results obtained with the improved and the standard period tests, we observed an overall power gain of ~20% for the *Z improved test* with respect to the *Z test* and ~50% in the case of *H improved test* (Caliandro 2006).

Further simulations have been performed in order to investigate the distributions of the best photon selection as a function of pulsar fluence. Several samples of 900 pulsars uniformly distributed along the galactic plane have been simulated, together with the galactic and extragalactic diffuse emission. Each sample is characterized by a value of fluence for *E*>100*MeV* which is attributed to all the pulsars of the sample.

We have chosen to simulate pulsars with double peak pulse shape, because all the known gamma-ray pulsars have this feature. The peaks have a Lorentian shape with random width and height; also the distance in phase of the two peaks is random, but it cannot be less than 0.3 phase. The spectrum of the pulsars is a cutoff-powerlaw (Nel, DeJager 1995):

$$\frac{dN_\gamma}{dE} = K\left(\frac{E}{E_n}\right)^{-g} \exp\left(\frac{-E}{E_0}\right)^{+b} \qquad (1)$$

where in our simulations we have set b=1, the break down energy E_0=10GeV and the spectral index g distributed as a Gaussian with mean μ=1.4 and σ=0.2.

The galactic diffuse emission is simulated following the map obtained from the *EGRET* data, instead the extragalactic emission is uniformly distributed in the sky.

Every pulsar has been analysed applying the *H improved test* and the cuts which give the best result have been recorded. In this way we had, for each sample of simulated pulsars, the distribution of the best cuts and it was possible to study the behaviour of the best cuts versus the fluence of the pulsars. Figures 3 and 4 show respectively the plots of the best *ROI* radius versus pulsar fluence and the best energy band low edge (E_{min}) versus pulsar fluence. In these plots the triangles correspond to the maximum of the respective distributions and the red bars are the spreads.

We can make some remarks from observing these plots. In the case one would use the original *H test* applying a single cuts, it is worth while to note that it is useful to apply ROI radius smaller than 2 degrees and an energy band low edge equal to or greater than 200MeV. In fact in figure 3, except for the lowest value of fluence, the peaks of the distributions are at 0.75° for low fluence and increase until 1.5° for higher fluence. Concerning the low edge of the best energy band (E_{min}) the distributions are peaked at values not less than 200MeV.

On the other hand, especially for low fluence the spreads of the distributions are very large; therefore it is not possible to define the best cuts as function of the pulsar fluence. For this reason the use of the *improved test*, instead of the original one, has to be preferred in order to increase the detection probability especially for low fluence pulsars.

4. Discussion

We have implemented an improvement for the period tests. The improvement consists in performing a period test for several different choices of photon selections and taking the best result among the values obtained.

Appling this improvement to several simulated pulsars, it results that the *Z improved test* has an overall increment in power of ~20% with respect the *Z test* and the *H improved test* an increment of ~50% with respect the *H test*.

The distribution under the null hypothesis of the improved tests is strongly affected by the preliminary choice of the photon selection criteria to investigate. For this work we have found this distribution simulating a huge sample of spurious sources.

A new Monte Carlo code is already being developed in order to shorten the time needed to find the distribution under the null hypothesis of the improved test. The code takes as input the definition of the photon selection criteria to investigate and returns the distribution under the null hypothesis using the statistical properties of the Z/H period tests, without simulating any spurious sources or diffuse emission. In this way for each set of photon selection criteria defined a priori we could have the respective distribution in a short time.

Acknowledgments

We would like to acknowledge David Smith, Jean Ballet, Dave Thompson and Patrick Nolan for their useful comments and suggestions.

References

1. Buccheri, R., et al., *Astron. Astrophys.* 128, 245 (1983)
2. Caliandro, G.A. doctoral thesis, University of Bari (2006)
3. De Jager, O.C., Swanepoel, J.W.H., Raubenheimer, B.C., *Astron. Astrophys.* 221, 180 (1989)
4. Leahy, D.A., et al., *Astrophys. J.* 266, 160 (1983a)
5. Leahy, D.A., et al., *Astrophys. J.* 272, 256 (1983b)
6. Nel, H. I., DeJager, O. C., Ap&SS 230, 299 (1995)

5. Figures

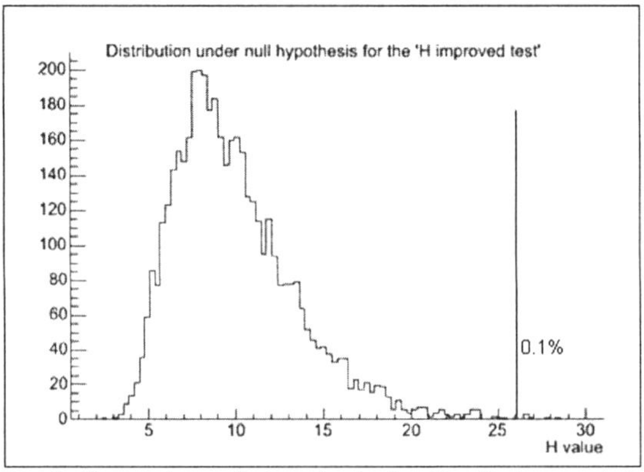

Fig1: Distribution under null hypothesis for the *H improved test*. This distribution is obtained applying the *H improved test* to a sample of 4500 spurious sources. The vertical line is the 0.1% significance threshold.

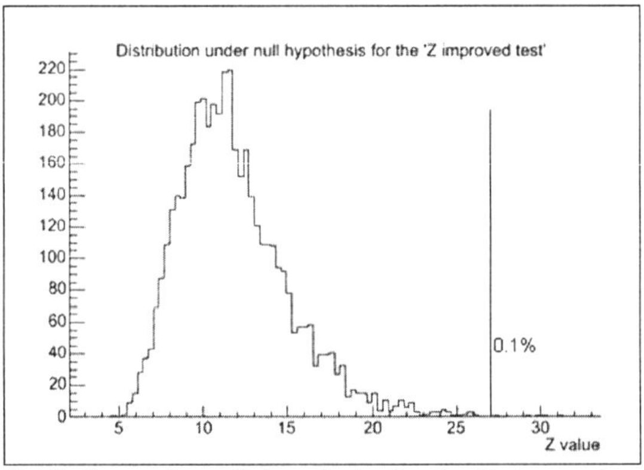

Fig2: Distribution under null hypothesis for the *Z improved test*. This distribution is obtained applying the *Z improved test* to a sample of 4500 spurious sources. The vertical line is the 0.1% significance threshold.

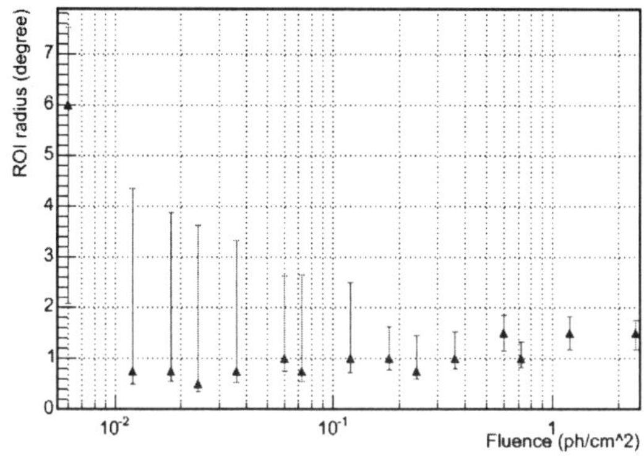

Fig3: Best *ROI* radius versus pulsar fluence.

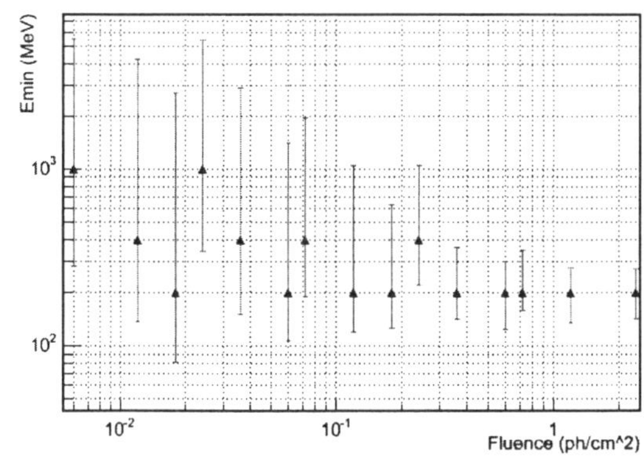

Fig4: Low edge of the best energy band (E_{min}) versus pulsar fluence.

Gamma-Ray Burst Physics with GLAST

Nicola Omodei

INFN of Pisa, Polo Fibonacci Largo B. Pontecorvo 3, Pisa, 56127, Italy

for the GLAST/LAT Collaboration

The Gamma-ray Large Area Space Telescope (GLAST), scheduled to be launched in late of 2007, is the next generation satellite for high-energy gamma-ray astronomy. The Large Area Telescope (LAT), the main instrument of GLAST, will survey the sky in the energy range between 20 MeV to more than 300 GeV, shedding light on many issues left open by its highly successful predecessor EGRET (Energetic Gamma Ray Experiment Telescope). The Gamma-Ray Bursts (GRBs) are one of the most exiting science topic for the LAT, which will open an new window, exploring these sources in a energy range never observed before. Even if little is know at high energy, we would like to address, in this contribution, some of the features of the GRB spectrum that theories predict and that LAT can observe. Finally we would like to present some results about the study of the sensitivity of the LAT to GRBs.

Keywords: Gamma-ray:bursts; Gamma-ray:telescopes

1. Introduction

GLAST[1] is an international mission that will study the gamma-ray Universe. It will be launched aboard a Delta 2920H-10 from Cape Canaveral, on a 565 km circular orbit at 25.3^o inclination. The heart of GLAST is the LAT, a pair production telescope sensitive to gamma rays in the energy range between 20 MeV-300 GeV and above. The LAT's energy range, effective area, field-of-view (FoV) and angular resolution are vastly improved in comparison with those of its highly successful predecessor EGRET (1991-2000), so that the LAT will provide a factor 30 or more advance in sensitivity[a]. This improvement should enable the detection of several thousands new high-energy sources and allow the study of GRBs and other transients, the search for dark matter, the detection of active galactic nuclei, pulsars,

[a]see http://www-glast.slac.stanford.edu/software/IS/glast_lat_performance.htm for the LAT performances

supernova remnants and the study of the extragalactic diffuse gamma-ray emission. The second detector onboard the GLAST satellite is the GLAST Burst Monitor (GBM),[2] which consists of 12 NaI detectors for the 10 keV to 1 MeV range and two BGO detectors for the 150 keV to 30 MeV range. It covers the entire visible sky not occulted by the Earth, extending the spectral coverage of GRBs down to the tens keV range, where most of GRB phenomenology is known since the BATSE instrument. For GRBs, GBM and LAT data will be jointly analyzed, providing information over more than seven energy decades.

2. Gamma-Ray Bursts and GLAST

Gamma-Ray Bursts are flashes of high-energy radiation, probably related to the explosions of massive stars or the merging of compact objects. Their energy is peaked, in the νF_{ν} spectra, at hundreds of keV, and is typically described with a smoothly broken power law.[13] GRBs are still a puzzling topic and few observations are currently available above 50 MeV. EGRET detected only a few high-energy bursts,[3] and did not detect the high energy cutoff or rolloff that must exist for some hard spectra to keep the energy flux finite. Surprisingly, GeV emission was found to last up to 90 minutes after the 150 keV emission;[4] in addition, an extra spectral component was observed at 100 MeV in GRB941017.[5]

During the first year it will operate in scanning mode, providing uniform sky coverage every three hours; nevertheless GLAST can re-point, keeping the burst in the LAT field of view studying the development of the burst emission, hours after the trigger, looking, eventually, for delayed high energy emission. GBM and the LAT can independently trigger on GRBs; a rapid alert message will be sent to the ground near real-time using the TDRESS satellites[b] providing basic information for follow-up observations. According to the predicted sensitivity (0.8 ph/cm^2/s), the GBM will detect ~200 bursts per year, of which more than 60 will fall in the 2.4 sr field of view of the LAT.[6] The initial on-board GBM localization accuracy is ~15 degrees (within 1.8 s), updates with better location will come later. The LAT detector can provide better accuracy, of the order of ten arc minutes or less, depending on the burst intensity. For a few bursts per year the LAT will localize them to sufficiently small error boxes (0.1 degrees) that medium field-of-view instruments can point for follow up observations, providing a

[b]A cluster of commercial telescopes, available for downlink the alert messages from GLAST.

more precise measurement of the GRB position. GLAST will be able to investigate the emission processes that are present in GRB phenomena and scan the most energetic region where particles are accelerated, probably reaching their highest energy.

3. GRB simulators

In order to simulate GLAST observation of Gamma-Ray Bursts, we adopt the the idea of developing a general interface to accommodate different models based on different approaches. In general we consider a GRB flux as a two dimensional histogram, where the flux (as $N_{ph}/cm^2/s$) is described as a function of energy and time. The physical model describe the GRB phenomena starting from the so called *fireball* scenario.[8] In this model we simulate the emission of shells of matter (e^+ and e^-, mainly), emitted by a central engine with relativistic velocities. The shells can collide producing shocks; part of the energy density dissipated during these shocks is converted in accelerated particles, and part goes in magnetic field energy density. In this standard scenario, particles cool by Synchrotron radiation whereas the pulse shape depends, basically, on geometrical shapes of the emitting shells. Fig.1 shows the typical spectral shape of a GRB, as described by the fireball model: the first peak of the νF_ν spectrum is due to Synchrotron radiation. Due to the slow cooling time, the synchrotron spectrum shows a cut-off at GeV energies, well in the LAT field of view. Natural extension at high energy is the Inverse Compton radiation, where the synchrotron photons are up-scattered by the accelerated electrons. This mechanism characterize the second bump in the schematic spectrum of Fig. 1. At high energy the IC spectrum is attenuated mainly by intrinsic absorption between IC photons and Synchrotron photons. This attenuation depends on the Lorentz factor of the emitted material and, for moderate Lorentz factor, can be in the LAT energy range.[18]

A different approach is followed in the phenomenological model, where the GRB behavior is based on observed quantities, mostly coming from the BATSE catalog (Preece et al. 2000[14]). In our model we obtain the temporal-spectral evolution of a GRB by multiplying the "Band" function[13] for a pulse shape, described by a universal family of functions. In agreement with the results from Fenimore et al. (1995)[11] and Norris et al.(1996)[9] the pulse width W depends on the energy ($W(e) = W_0(e/20keV)^{-0.33}$); the spectrum is extrapolated up to LAT energies as well as pulse width scale law. Thus, this model foresee narrower pulses at high energy, as narrow as millisecond time scale. This model, firstly developed by Norris et al.(1996)[9]

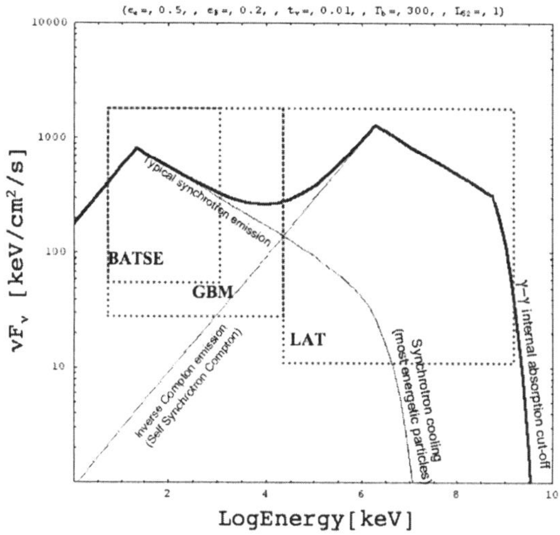

Fig. 1. Schematic representation of the spectrum of a GRB as described by the fireball model. In this model we have included the synchrotron spectrum, and the Inverse Compton Component. The syncrotron spectrum is limited at high energy by the maximum energy reached by the accelerated particles, while the Inverse Compton is limited by intrinsic absorption (we have approximated the two components by broken power laws. A detailed calculation would have result in a smoother spectrum. In the pictures the the energy band for the BATSE, GBM and LAT detectors are depicted.

has been recoded and it is now accessible as part of the *Science Tools.* [c] For each simulated burst a sequence of photons is sampled, in agreement with the spectral-temporal development of the flux. Each photon is than processed by a Monte Carlo simulator, based on GEANT4 toolkit, which follows the propagation of particles in the detector. Another (faster) approach is to use the parameterized response of the instrument, using the Instrument Response Function (IRF). Even in this case the followed approach is photon-by-photon, and the output format is the same as in the case of the full Monte Carlo approach. GRBs can be combined with other classes of sources, (stationary and flaring AGN, solar flares, SNR, Pulsars,...), building a very complex and, possibly realistic, picture of the gamma-ray sky. The GRB simulators are also able to interface the GBM software providing correct input files for the detector simulators. In this case the approach is slightly different. The GRB simulators produce a temporal series of Band parameters, fitting continuously the GBM counterpart of a LAT bursts. These parameters are then used to compute the GBM response and the final output is a series of *FITS* files, that can be analyzed jointly with LAT data, producing sky-maps, lightcurve, and spectral joint spectral fits. Fig.2 is an example of combined light curve of a simulated bursts. Combining GBM and LAT data a GRB can be simultaneously studied in an energy

[c] A set of tools for analyzing (and simulating) LAT data, that the LAT team will deliver to the community.

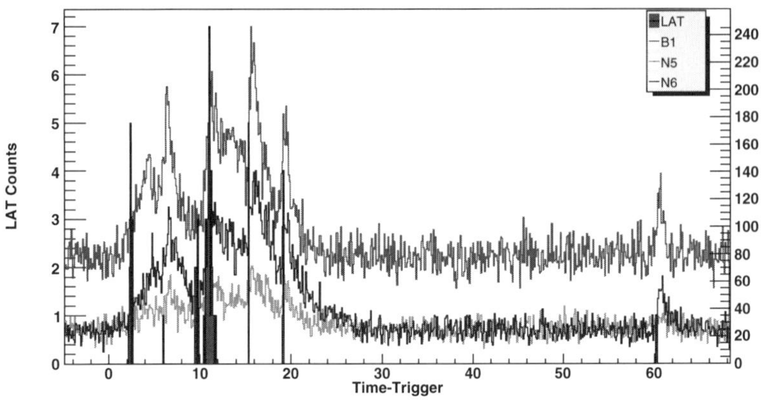

Fig. 2. Combination of the simulated counts rate of the BGO (B1), two NaI detectors (N5 and N6) of the GBM experiment, over-imposed with the LAT light curve. LAT will have enough temporal resolution to correlate low energy spikes with high energy temporal profiles, understanding details of the emission process. In the figure is well visible the 'time-lag' between low energy and high energy temporal structures as well as the characteristic 'pulse paradigm', for which high energy pulses are narrower than low energy pulses.

band larger than seven orders of magnitude, making GLAST a very powerful tool for understanding the correlation between the low-energy and high-energy spectra in GRBs.

4. Study the LAT's GRB sensitivity

We used the BATSE catalog for building up our statistics. Considering the flux threshold (0.3 ph/cm^2/s) and the field of view of the BATSE instrument, the expected number of bursts per year over the full sky is 650. For each simulated burst the duration is drawn from the observed T_{90} distribution and its flux is sampled from the BATSE peak flux distribution in the 50-300 keV energy range.[12] The number of pulses is fixed by the total burst duration. The peak energy e_p and the low and high energy spectral indices α and β are sampled from the observed distributions.[13,14] In this model, long and short bursts are treated separately. At high-energy photons produced by GRB at cosmological distances can be absorbed by the Extragalactic Background Light (optical-UV photon), producing pairs. In our simulation we have included this effect, adopting the EBL model proposed in[15] and adopting the redshift distribution for short bursts proposed by Guetta et al.(2005).[16] Long bursts are likely related to the explosive end of massive stars, whose distributions are well traced by the Star Formation

History.[17]

We simulate one year of observations in scanning mode. The orbit of the GLAST satellite, the South Atlantic Anomaly (SAA) passages and the Earth occultation are correctly considered. In Fig.3 we plot the number of expected bursts per year as a function of the number of photons per burst detected by the LAT, where different lines refer to different energy thresholds. The EBL attenuation affects only the high-energy curve, as expected from the theory, leaving the sensitivities almost unchanged with thresholds less than 10 GeV. In this calculation, LAT will independently detect 50-70 burst per year, depending on the sensitivity of the detection algorithm. Few bursts per year will have a sufficient amount of counts in the LAT detector to allow detailed spectra studies, combining the LAT and the GBM spectra.

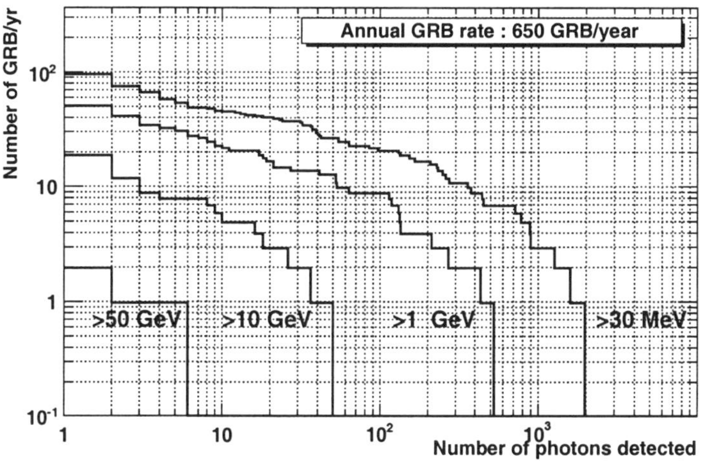

Fig. 3. Model-dependent LAT GRB sensitivity. The GRB spectrum is extrapolated from BATSE to LAT energies. The burst rate in the 4π sphere is assumed to be 650 GRB/yr (above 0.3 ph s^{-1} cm^{-2} in the 50-300 keV band), in agreement with BATSE statistics. The bursts here simulated do not have the Inverse Compton Component. The effect of the EBL absorption is included. Different curves refer to different energy thresholds.

5. Conclusions

LAT is a new window to GRBs, covering energy where we basically don't have information about how the GRB spectrum looks like. Different emission mechanism can take place at high energy; the most probable one is the Inverse Compton scattering. This component can enhance the spectrum at the LAT energies increasing the GRB yield. On the other hand, absorption phenomena can suppress the spectrum in the LAT domain. In case of IC component, the intrinsic absorption by pair production is the most effective one, and it is the limit for high energy emission. In case of moderate bulk Lorentz factor of the emitting shell, this component is in the LAT energy range, and in case of bright burst can be measured with statistically significance.[18] Further works will study these effect in detail.

References

1. P. F. Michelson, *The Gamma-ray Large Area Space Telescope Mission:Science Opportunities*, in *AIP Conf. Proc.* **587**, 713–+ (2001).
2. A. von Kienlin et al., *The GLAST Burst Monitor for GLAST* in *Proc of the SPIE-Conference* (2004)
3. B. L. Dingus, *Astrophys. & Space Sci.* **231**, 187–190 (1995).
4. K. Hurley et al., *Bulletin of the American Astronomical Society* **26**, 881–+ (1994).
5. M. M. González et al., *Nature* **424**, 749–751 (2003).
6. D. L. Band, *Ap. J.* **588**, 945 (2003) [arXiv:astro-ph/0212452].
7. N. Gehrels et al. Ap. J., 611, 1005 (2004)
8. T. Piran, Physics Reports, **314**, 57 (1999)
9. J. P. Norris et al., *Ap. J.* **459**, 393–+ (1996).
10. S. P. Davis, at al.,*Pulse Width Distributions and Total Counts as Indicators of Cosmological Time Dilation in Gamma-Ray Bursts*, in *AIP Conf. Proc.***307**, 182–+, (1994).
11. E. E. Fenimore, et al. Ap. J. Lett., **448**, L101, (1995).
12. W. S. Paciesas *et al.*, Astrophys. J. Suppl. **122**, 465 (1999) [arXiv:astro-ph/9903205].
13. D. Band et al., *Ap. J.* **413**, 281–292 (1993).
14. Preece et al., *Ap. J. Supp.* **126**, 19–36 (2000).
15. J. R. Primack, J. S. Bullock, and R. S. Somerville, *Observational Gamma-ray Cosmology*, in *AIP Conf. Proc.* **745**, 23–33 (2005).
16. D. Guetta, and T. Piran, *Astron. & Astrophys.* **435**, 421–426 (2005).
17. C. Porciani, and P. Madau, *Ap. J.* **548**, 522–531 (2001).
18. M. G. Baring, *Ap. J.* **650**,1004 (2006)

The Global fit Approach to time-resolved spectroscopy of GRBs

Anton Chernenko*

Space Research Institute,
Moscow, Russia
* *E-mail: anton@cgrsmx.iki.rssi.ru*
www.iki.rssi.ru

We describe a general method for time resolved gamma-ray spectroscopy of astrophysical transients - the Global Fit Analysis. Instead of parameterizing individual spectra and analyzing spectral evolution of a transient in terms of numerous individual spectral fit parameters we define a spectral evolution model with a set of constant global parameters and just a few (N=1-2) time dependent variables.

In case of GRBs, Global fits explain spectral variability in terms of single variable component with much simpler properties than normally assumed for GRB emission, being are statistically as good as traditional bin-to-bin fits.

Keywords: gamma-rays, spectroscopy, optimization, GRB.

1. Definition of the Global Fit Analysis

The core of the method is the following postulate: spectral variability of a given gamma-ray burst is completely determined by 2 variable parameters: $X(t)$ and $Y(t)$ and a set of constant parameters $\{\Omega\}$. The selection X, Y and $\{\Omega\}$ depends on a particular kind of model function used in a particular type of the Global fit analysis.

The total number of adjustable parameters is too large for them to be estimated via a usual optimization procedure. Thus, we were forced to implement a special two-level procedure that is outlined below.

(1) Given a trial $\{\Omega\}$ we find values X_j, Y_j for every time interval j that minimize the values of χ_j^2 that is a sum of normalized squared differences between the j-th data column (spectrum) and the corresponding spectral function $S_{ij} = S(E_i; X_j, Y_j, \{\Omega\})$:

$$\chi_j^2 = \frac{(D_{ij} - S_{ij})^2}{\delta D_{ij}} \tag{1}$$

(2) Then the optimal set $\{\Omega\}$ is found by minimizing the global χ^2 value

$$\chi^2 = \sum_j \chi_j^2 \tag{2}$$

As a result, one obtains both time histories of the parameters $X(t)$, $Y(t)$ and the set of best fit Global parameters $\{\Omega\}$. While $X(t)$, $Y(t)$ characterize internal properties of a given burst, the set $\{\Omega\}$ could be used for comparing different bursts and their classification.

So far, we have successfully applied and tested 2 models used for global fits: two component model (one variable parameter per component) and single component model (two variable parameters). The two component model was presented in Chernenko (2002) [1] and this paper will concentrate on the single component model.

2. Single component model

In this type of model we postulate here that the photon spectrum $S(E, t)$ of GRB emission at any moment t can be represented as a single component with the spectral shape proposed by Band et al. (1993) [2].

$$\begin{aligned} S(E) = {} & AE^{\alpha} \exp[-E/E_0], && E < E_m \\ & A[(\alpha - \beta)E_0]^{\alpha-\beta} \exp(\beta - \alpha)E^{\beta}, && E > E_m \end{aligned} \tag{3}$$

where $E_m = E_0(\alpha - \beta)$. Then, each of the spectral parameters α, β, and E_0 is expressed in terms of some 2 parameters that correspond to this time interval. For computational reasons we select logarithm of apparent count rate $lg(R)$ and $lg(E_0)$ as such independent parameters. In this case the other parameters are expressed in terms of these two as follows:

$$\alpha = \alpha_0 + \alpha_{\mathrm{R}} \lg R + \alpha_{\mathrm{E}} \lg E_0 \tag{4}$$

$$\beta = \beta_0 + \beta_{\mathrm{R}} \lg R + \beta_{\mathrm{E}} \lg E_0 \tag{5}$$

Therefore, the set of Global parameters is as follows: $\Omega = \{\alpha_0, \alpha_{\mathrm{R}}, \alpha_{\mathrm{E}}, \beta_0, \beta_{\mathrm{R}}, \beta_{\mathrm{E}}\}$.

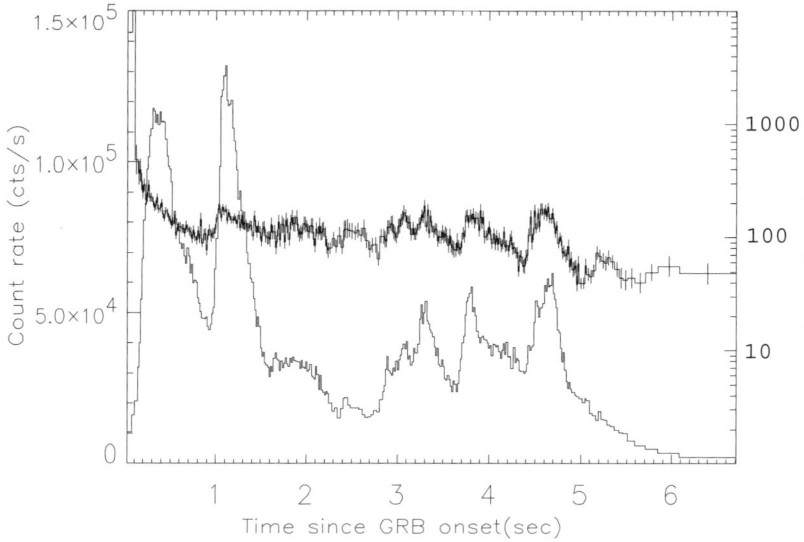

Fig. 1. Time history of the count rate for GRB930922. Also shown is time history of break energy E_0 (keV) estimated via the Global fit.

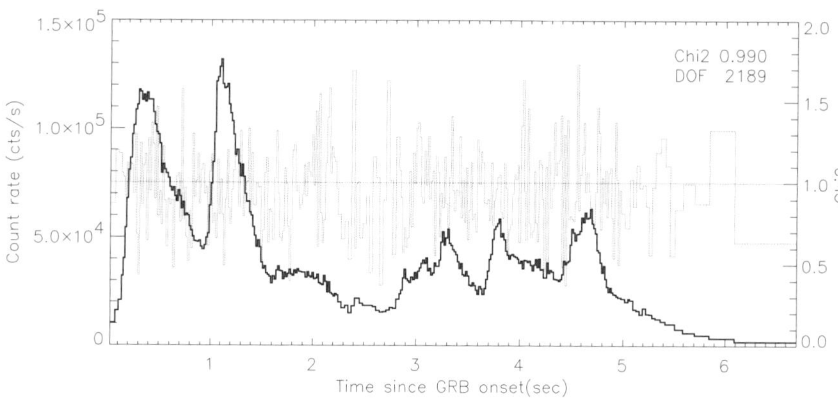

Fig. 2. The time history of χ_j^2 (see eq. 1) is plotted (dotted line) over the time history of count rate of GRB930922. Also indicated is the reduced Global fit χ^2 (see eq. 2) and the total number of DOF.

In Figure 1 the Global fit is presented for GRB 930922, as recorded by BATSE/CGRO experiment in the energy range of 25–2000 keV. These two variable parameters - count rate and the break energy E_0, together with the estimated set Ω completely explain spectral variability of the GRB.

The time history of reduced χ_j^2 (eq. 1) presented in Figure 2 by dotted line shows no apparent deviation from 1.0 and manifests no temporal structure. The reduced Global $\chi^2 = 0.99$ (eq. 2) also indicates that the Global fit for this GRB is statistically acceptable.

Application of the Global fit analysis to the brightest GRBs from BATSE data is summarized in Table 1. Except the sole case of GRB 940703 it was confirmed that the Global fits were statistically as good as traditional bin-by-bin fits. It is expected and will be verified that for fainter GRBs with worse S/N ratio this finding will be confirmed.

Table 1. Statistical comparison of Global and traditional fits for the brightest GRBs recorded by BATSE/CGRO experiment

Trigger	GB Name	Global Fit		Traditional Fit		P(F-test)
		χ^2_{glob}	DOF_{glob}	χ^2	DOF	
143	910503	1.063	4878	1.031	3911	0.002
249	910601	1.049	11480	1.048	9046	0.420
1085	911118	1.026	6921	1.091	5133	1.000
2083	921207	1.008	4075	1.089	2750	1.000
2151	930131	1.380	354	1.340	278	0.210
2329	930506	1.080	10464	1.070	8216	0.100
2537	930922	0.991	2185	1.042	1571	0.990
2798	940206	0.997	11662	1.032	9172	1.000
2953	940429	1.017	4034	1.030	3174	0.740
3057	940703	1.360	8500	1.160	6660	0.000
3765	950818	1.070	1878	1.060	1303	0.390
3492	950403	1.120	3147	1.190	2420	1.000
3523	950425	1.010	11285	1.090	8859	1.000
5989	970201	1.280	781	1.310	615	0.870
6124	970315	1.110	5907	1.097	4543	0.140
6198	970420	1.065	5236	1.090	3996	0.990
6587	980203	1.061	10460	1.046	8180	0.020
7301	990104	1.062	4585	1.047	3519	0.100

References

1. A. Chernenko *in the Proceedings of the IAU 8th Asian-Pacific Regional meeting*, v.2, eds. S. Ikeuchi, J. Hearnshaw and T. Hanawa, p.321 (2002)
2. D. L. Band, et al., *ApJ*, **413**, 281 (1993).